SYMBOLIC COMPUTATION
IN
UNDERGRADUATE MATHEMATICS
EDUCATION

SYMBOLIC COMPUTATION
IN
UNDERGRADUATE MATHEMATICS
EDUCATION

Zaven A. Karian, Editor

The Mathematical Association of America

MAA Notes and Reports Series

The MAA Notes and Reports Series, started in 1982, addresses a broad range of topics and themes of interest to all who are involved with undergraduate mathematics. The volumes in this series are readable, informative, and useful, and help the mathematical community keep up with developments of importance to mathematics.

MAA Notes

1. Problem Solving in the Mathematics Curriculum,
    Committee on the Teaching of Undergraduate Mathematics,
    a subcommittee of the Committee on the Undergraduate Program in Mathematics, *Alan H. Schoenfeld,* Editor.

2. Recommendations on the Mathematical Preparation of Teachers,
    Committee on the Undergraduate Program in Mathematics, Panel on Teacher Training.

3. Undergraduate Mathematics Education in the People's Republic of China,
    Lynn A. Steen, Editor.

4. Notes on Primality Testing and Factoring,
    Carl Pomerance.

5. American Perspectives on the Fifth International Congress on Mathematical Education,
    Warren Page, Editor.

6. Toward a Lean and Lively Calculus,
    Ronald G. Douglas, Editor.

7. Undergraduate Programs in the Mathematical and Computer Sciences: 1985–86,
    D. J. Albers, R. D. Anderson, D. O. Loftsgaarden, Editors.

8. Calculus for a New Century,
    Lynn A. Steen, Editor.

9. Computers and Mathematics: The Use of Computers in Undergraduate Instruction,
    Committee on Computers in Mathematics Education, D. A. Smith, G. J. Porter, L. C. Leinbach, and R. H. Wenger, Editors.

10. Guidelines for the Continuing Mathematical Education of Teachers,
    Committee on the Mathematical Education of Teachers.

11. Keys to Improved Instruction by Teaching Assistants and Part-Time Instructors,
    Committee on Teaching Assistants and Part-Time Instructors, Bettye Anne Case, Editor.

12. The Use of Calculators in the Standardized Testing of Mathematics,
    John Kenelly, Editor, published jointly with The College Board.

13. Reshaping College Mathematics,
    Committee on the Undergraduate Program in Mathematics, Lynn A. Steen, Editor.

14. Mathematical Writing,
    by *Donald E. Knuth, Tracy Larrabee, and Paul M. Roberts.*

15. Discrete Mathematics in the First Two Years,
    Anthony Ralston, Editor.

16. Using Writing to Teach Mathematics,
    Andrew Sterrett, Editor.

17. Priming the Calculus Pump: Innovations and Resources,
    Committee on Calculus Reform and the First Two Years,
    a subcommittee of the Committee on the Undergraduate Program in Mathematics, *Thomas W. Tucker,* Editor.

18. Models for Undergraduate Research in Mathematics,
Lester Senechal, Editor.

19. Visualization in Teaching and Learning Mathematics,
Committee on Computers in Mathematics Education, Steve Cunningham and Walter S. Zimmermann, Editors.

20. The Laboratory Approach to Teaching Calculus,
L. Carl Leinbach et al., Editors.

21. Perspectives on Contemporary Statistics,
David C. Hoaglin and David S. Moore, Editors.

22. Heeding the Call for Change: Suggestions for Curricular Action,
Lynn A. Steen, Editor.

23. Statistical Abstract of Undergraduate Programs in the Mathematical Sciences and Computer Science in the United States: 1990–1991 CBMS Survey,
Donald J. Albers, Don O. Loftsgaarden, Donald C.Rung, and Ann E. Watkins.

24. Symbolic Computation in Undergraduate Mathematics Education,
Zaven A. Karian, Editor.

MAA Reports

1. A Curriculum in Flux: Mathematics at Two-Year Colleges,
Subcommittee on Mathematics Curriculum at Two-Year Colleges, a joint committee of the MAA and the American Mathematical Association of Two-Year Colleges, *Ronald M. Davis,* Editor.

2. A Source Book for College Mathematics Teaching,
Committee on the Teaching of Undergraduate Mathematics, Alan H. Schoenfeld, Editor.

3. A Call for Change: Recommendations for the Mathematical Preparation of Teachers of Mathematics,
Committee on the Mathematical Education of Teachers, James R. C. Leitzel, Editor.

4. Library Recommendations for Undergraduate Mathematics,
CUPM ad hoc Subcommittee, Lynn A. Steen, Editor.

5. Two-Year College Mathematics Library Recommendations,
CUPM ad hoc Subcommittee, Lynn A. Steen, Editor.

First Printing
© 1992 by the Mathematical Association of America
ISBN 0-88385-082-6
Library of Congress Catalog Card Number 92-64063
Printed in the United States of America

Preface

Although symbolic computing systems have been available for about 25 years, their use in mathematics education has been a relatively recent phenomenon. By current standards, the early versions of these systems were cumbersome to use and were quite expensive. In 1984 a single user system (hardware and software) dedicated to symbolic computation cost about $40,000; today more sophisticated systems can be obtained for less than $2,000. The general availability of these powerful and inexpensive systems has the potential of revolutionizing both mathematics instruction and the practice of mathematics.

St. Olaf College began experimenting with using symbolic computation in calculus instruction in 1986; Colby College initiated a similar effort at about the same time. Within a couple of years, at least 20 institutions were using symbolic computation in some part of their mathematics curriculum. These initial efforts coincided with the concern about reforming calculus instruction in the nation's colleges and universities. The 1986 Tulane Conference (for the proceedings of this conference see R. G. Douglas (ed.), *Toward a Lean and Lively Calculus*, MAA Notes No. 6) firmly established calculus reform as a major concern of mathematics education.

Two papers presented at the Tulane Conference, one by Don Small and John Hosack, "Computer Algebra Systems: Tools for Reforming Calculus Instruction," and the other by Paul Zorn, "Computer Symbolic Manipulations in Elementary Calculus," as well as issues raised by other papers (e.g., Should Calculus be a Laboratory Course, a question raised by Lynn Steen in his "Twenty Questions for Calculus Reforms") developed a connection between calculus reform and the use of symbolic computation in calculus instruction. This connection was strengthened a year later when the Sloan Foundation, the sponsor of the Tulane Conference, provided support to eight institutions (Colby College, Denison University, Harvey Mudd College, Oberlin College, Rollins College, University of Hawaii, University of Saskatchewan, University of Waterloo) to investigate the improvement of calculus instruction through the use of symbolic computation. Following the Sloan Foundation initiative, about a dozen grants were made by the CSIP (the precursor of ILI) program of NSF to support the acquisition of computing equipment.

In October of 1988, the MAA decided that "the challenge to the curriculum posed by symbolic computer systems was so special and potentially so powerful" that it merited national attention. Accordingly, the MAA established a Subcommittee of CUPM, the Subcommittee on Symbolic Computation (SSC), with the principal charge: "to coordinate the existing disparate activities in symbolic computation, to promote new initiatives, and to disseminate the work being done in this area to the larger mathematics and science community." Shortly after its formation, the SSC obtained funding from the Sloan Foundation to support its activities.

Since 1988, the number of faculty involved with using symbolic computation has increased dramatically. Two of the major contributors to this have been the activities sponsored by the SSC and the NSF-funded workshops organized by Don Small.

For many of us who monitor curricular changes semester by semester, pedagogic initiatives seem slow to catch on. But, viewed over a period of five years, the impact of symbolic computation on the undergraduate curriculum is quite impressive. This volume, developed under the aegis of SSC, brings together many of the facets associated with the pedagogic uses of symbolic computation. The articles are clustered by the areas of the curriculum they address and not by the particular symbolic computing system that they use. The particular technology employed varies from HP28S calculators to *Derive*, *Mathematica*, and *Maple*.

Part I consists of articles that deal with general issues of learning mathematics and the role of symbolic computation in that process. During the early years (1986–1988), the use of symbolic computation in mathematics education was closely coupled with the calculus reform movement. Consequently, more attempts have been made to use symbolic computation in calculus than in other parts of the mathematics curriculum. The papers in Part II describe some of the experiences of our colleagues in this area.

Linear algebra and differential equations are considered sophomore-level courses at most institutions. Most mathematics departments teach "stand-alone" courses in these areas, others teach a combined linear algebra and differential equations course, and a few teach linear algebra in conjunction with calculus. Part III consists of papers on linear algebra and differential equations taught in a variety of settings.

The use of symbolic computation has moved to the more advanced undergraduate mathematics courses in the last couple of years. The two articles in Part IV describe what can be done in probability and statistics and combinatorics courses. The articles in Part V provide information and assistance to novices who are considering incorporating symbolic computation into their curriculum. The first article relates the experiences at St. Olaf College as it got involved with symbolic computation. The other two articles give an annotated guide to the literature.

I have been fortunate to receive the assistance of many colleagues in the preparation of this volume. Wade Ellis, Robert Lopez, Arnold Ostebee, Stanley Seltzer, Donald Small, David Smith, and Warren Page all reviewed portions of the submitted manuscripts and made valuable suggestions. I am particularly indebted to Andrew Sterrett, who helped me with a second review by reading and commenting on virtually every article that was submitted.

Zaven A. Karian
Denison University
May, 1992

CONTENTS

Part I:

General Pedagogic Issues

Questions for the Future: What About the Horse? *

David A. Smith
Duke University

1 The Big Question

What are the questions for the future regarding the tough issues and pressures we will face in calculus reform as a consequence of the emergence of computer algebra systems (CAS's)? Tempted to make predictions, I am reminded of a line penned by Jonathan Swift: "Old men and comets have been reverenced for the same reason, their long beards and pretences to foretell events."[1]

I have about the same chance as Halley (the comet, not the man) of predicting what will happen to the curriculum when would-be reformers confront the educational establishment. Thus, I will let others make the predictions. Quoth the Oracle:

> The availability of [CAS's] has created new challenges for curriculum development in mathematics. Specifically, the undergraduate mathematics curriculum must be changed to take into account new ways of solving old problems and the host of new problems which have arisen. Students will be enormously stimulated by the introduction of [CAS's] into the curriculum; for it will turn the all-too-often totally passive college experience into one of active participation.

> If the basic undergraduate mathematics courses are not appropriately modified to reflect the points of view which are associated with [CAS] applications in mathematics, these courses will lose much of their relevance for the coming generation of college and university students.

Pop quiz: Who said that, and when? It could have been said by almost anyone at any of the recent symposia, committee meetings, workshops, minicourses, contributed paper sessions, or conferences on CAS. This is a trick question because, in the brackets, I replaced the original text with "CAS's," our current fascination. The original predates CAS's as we know them; it is taken from the *Proceedings of a Conference on Computers in Undergraduate Education: Mathematics, Physics, Statistics, and Chemistry*, sponsored by the National Science Foundation at the University of Maryland in December, 1967.[2]

And what happened to the five or six college generations who have come and gone since that ringing pronouncement?

> Calculators and computers have had virtually no impact on mathematics instruction in spite of their great potential to enrich, enlighten, and expand students' learning of mathematics.
> *Everybody Counts*, [3]

> We have drifted into a curriculum by default, a curriculum of minimum expectations that resists the changes needed to keep pace with the demands of preparing students for contemporary life.
> John A. Dossey, quoted in [3], *Everybody Counts*

What went wrong? It has often been suggested that "technology drives the curriculum," and there is

---

*This paper is based on a talk given at the St. Olaf Conference on Symbolic Computing Systems, Oct. 20–22, 1989.

[1] *Thoughts on Various Subjects, Moral and Diverting* (1706).

[2] Quoted (correctly) in Porter [5].

1

some truth in that. It's seldom the other way around; that is, the development of new technology is seldom in response to perceived needs in education, but rather in response to needs of sectors better able to pay. However, computer and calculator technologies have **failed** to move the curriculum out of the drifting course described by Dossey. Do we really think, now that our computer technology has evolved to another level, that **that** will make the difference?

There is one rather limited sense in which technology **should** drive the curriculum. Courses in mathematics that ignore the impact of technology on present and future practice of science, engineering, and mathematics perpetrate a fraud upon our students. In the world our students will inhabit, the successful individual will live by the dictum, "Machines compute; people think." Thus, we cannot afford to continue programming our students to function as inferior machines.

> Technology should be used not because it is seductive, but because it can enhance mathematical learning by **extending each student's mathematical power**. Calculators and computers are not substitutes for hard work or precise thinking, but challenging tools to be used for productive ends.
> *Everybody Counts*,[3], (emphasis added)

What are our productive ends? As an educator, my highest goals are for my students to **learn** and to **learn to learn**. Neither of these is equivalent to **good teaching**. Indeed, Schoenfeld (see [7]) has argued that instructor and student performance that is measured as "successful" may actually impede the acquisition and use of mathematical knowledge! I am coming to see my proper role much less as "instructor" or "lecturer" or "transmitter of knowledge" and much more as facilitator, enabler, guide, supporter, goad, and—perhaps most important—fellow learner in a community of learners. Computer tools can help me in all those roles, but not if I use them for flashier "show and tell."

Our fundamental problem for curriculum reform is a classic "cart before the horse" dilemma. "Technology drives the curriculum" suggests not only that the cart of technology is in front of the horse of education (as expressed through curriculum) but that it actually provides the motive power! In my view, the most important question for curricular reform is, "Never mind the cart; what are we going to do about the horse?"

2 Computer Algebra Systems

This essay is about both the cart and the horse, as well as the linkage between them. Therefore, I have to say a few words about CAS's—what they are, where they fit into the picture, what they might become.

What computer systems are we talking about? Schoenfeld (see [6]) offers the following (overlapping) classification of computer systems with real or potential benefits for education: drill-and-practice environments, tools that do the drudgework (so you can use your brains), gaming environments, simulations, dynamic representations, programming environments (non-standard consequences thereof), intelligent tutoring systems, microworlds, and transcendent technologies (those that allow us to do things no one could do before, at any speed). He offers mathematical examples, many of which are exciting, in each of these categories except the last: Enthusiasts of *Mathematica* notwithstanding, there are no transcendent technologies yet for mathematics. Where do CAS's fit? As an example in the "tools" category, along with the *Geometric Supposers*.

We might like to think of CAS's as our first example of a transcendent technology, but we know better; by themselves, these systems will not fundamentally alter the way we think about, do, or teach mathematics in the same sense that, say, the word processor (including its various extensions: outliners, idea managers, spelling checkers) has fundamentally altered the way we think about, do, and teach writing. On the other hand, we see in the current generation of CAS's a **potential** for the emergence of such a technology.

For purposes of this essay, I see a computer algebra system as (at least) an integrated symbolic, numeric, and graphical system with interactive and procedural interfaces. I would like to imagine, but cannot picture clearly, another generation of systems that include (transparently and without burdensome overhead) many other features: technical word processing, symbolic input[3] as well as output, simulations, dynamic representations, hypertext, and interfaces to videodiscs, databases, satellite communications, and supercomputing.

---

[3] I mean by this some form of liberation from the constraints of the typewriter keyboard, however enhanced it may become. Some such technologies already exist: dragging symbols with a mouse (too awkward) or dedicated keys for symbols (such as the [integral] and [derivative] keys on the HP-28S calculator—but there are too many symbols for any fixed keyboard). A better solution might be LCD or LED keycaps that would show new symbols each time the keyboard is reprogrammed.

Alvarado and Ray [1], suggest an "evolutionary" classification of computer uses in education that consists of three "eras": the programming era (in which students and instructors necessarily function as programmers), the packaged software era (in which students and teachers share the role of users, but are distinct from programmers), and the era of declarative languages (in which teachers can assume a distinct role as creators of environments for students, while still remaining distinct from programmers). The eras overlap, of course; all three paradigms are in use today. The declarative mathematical paradigm ideally is able to comprehend all types of mathematical constructs and relationships. In addition, it should have interactive access to the features we associate with CAS's, to editing and word processing, to a rich array of symbolic transformations. In short, it should be possible to "speak mathematics" to the computer, to tell the machine what we want done, and expect that it will be done.

As this is written, Soft Warehouse has just released Version 2.0 of *Derive*, the first version to have any programming capability at all.

Most of the CAS's we know today fit into this declarative paradigm. The cited paper is support for Alvarado's own system, SOLVER-Q, which includes, for example, a geometric modelling capability. Another system, CAL (White, [9]), is designed for the instructor to construct interactive learning environments with a built-in word processor to which student-users also have access; the "notebook" concept in *Mathematica* offers a similar capability. Each of these has many of the features described by Alvarado and Ray, but, as is the case with most extant CAS's, learning to use them effectively is far from transparent. The most transparent CAS is almost certainly *Derive*, but it sits firmly in the "tool" category, lacking a real procedural interface or capability for creating student learning environments. Our transcendent technology has not yet arrived.

3 Issues for the Future

Warren Page (see [4]) has discussed at some length the teaching and learning issues that result from introduction of CAS's. I take the liberty of summarizing some of his key points, even though I have too little space to do his essay justice.

Page believes that the most critical factors in teaching mathematics are "what" we convey to students (both consciously and unconsciously, and including the nature and value of mathematics) and "how" that communication takes place (including our actions that imprint concepts of what it means to "do mathematics" and who should do it). His "compelling questions" are:

1. How can CAS's be used to provide students with greater understanding and deeper insights than heretofore possible?

2. What new awareness and knowledge do we need in order to effectively harness and realize CAS's' great potential for educational gain?

3. Are we prepared to make the necessary commitments to the demands of that new awareness?

Page cites literature that illustrates how CAS's can enable students to comprehend rich structures, visualize complicated relationships, increase "malleability" of objects by varying parameters, pursue limiting cases, and make intuitive leaps.

On the other hand, through a series of interesting examples, Page attempts to show a variety of ways in which CAS's can "constrain, misdirect, or in other ways influence how students think and learn mathematics." He writes:

> CAS's can vitiate conceptual understanding if they are permitted to foster an immediacy toward, or an overemphasis on, computation. CAS's can also preclude or thwart creative thinking if we allow them to anesthetize our impulses to consider other representations, to seek new relationships lurking in representations, and to be innovative in how we process information.

But, to be honest, the non-CAS ways Page cites for looking at interesting problems are, for the most part, not things any of our students have ever seen, with or without our help. The possible evils he sees in the use (or abuse) of CAS technology are already evils in our educational system, even without technology. Conventional teaching of mathematics already anesthetizes whatever impulses students might have to vary the representations of their problems or to process information in any innovative way. Whether we are discussing blackboards or computers, it's not the tools that create these distortions of education. The real threat posed by availability of more powerful tools is that they will enable the educational establishment to scale new heights in its lemming-like drive to replace education with training—that they will reinforce old patterns instead of sweeping them away.

The flip side of Page's message I can easily endorse: "Neither CAS's, nor any other prescribed representational medium, can be ... the sole means to accommodate the rich and diverse ways we process information, formulate conjectures and attempt to solve problems." Indeed, while the thrust of his examples is to show that important and interesting representations of problems arise in ways not easily accommodated by (existing) CAS's, the examples also show that symbolic systems enable representations we couldn't have so easily in any other way. We come back to the question: Can we allow the cart to determine where we will go with our horse?

There is much more in Page's essay, including the potential impact of CAS's on teaching and learning environments, the faculty, symbol sense and its acquisition, verification of solutions, formulation of conjectures, exploration, syllabus topics, the paucity of research on how students learn, and the role of assessment. With regard to the last, he writes

> We stress enhanced understanding and critical thinking, but our test questions stress routine processes and computations exemplified by the verbs 'solve', 'sketch', 'evaluate', 'calculate', 'compute', 'differentiate', 'integrate', 'invert', etc. ... precisely those mindless processes keyed to what CAS's can do best. ... Asking students to 'state and prove' also reveals no understanding of what may have been memorized. A first step toward reducing the disparity between our educational goals and our assessment objectives is to begin to use examination questions characterized by the verbs 'define and illustrate', 'explain', 'express', 'describe', 'compare', 'justify', 'interpret', etc. ... actions that transcend or envelop first-order CAS processing. ... Fifteen or twenty minutes of probing discourse with a student, using a CAS, could reveal a great deal (to the student as well as the instructor) about the student's state of knowledge ...

4   Questions and Answers

Others (e.g., Hosack, [2]) have compiled lists of questions, issues, potential problems that are raised by

the very existence and presumed availability of CAS's. Warren Page and I compiled such a list, when we were both involved with *The College Mathematics Journal*, for a "Forum" that never materialized but those discussions became part of the background of the paper cited above. I close with a list of those questions that appear important to me today; for some of these questions, I speculate on what the answers might be.

Teaching. What courses in the curriculum are or should be or will be affected? Who is affected, and in what way? Are we teaching the wrong skills and concepts for the future? What will be the new demands on teachers and the new dimensions of teaching? Will teachers have to become facilitators, choreographers, orchestrators of learning?

Learning. What do we really know about how students learn? What is the role of hand computation in that learning? Will the presence of CAS's erode learning of important skills? Will students be encouraged to compute first, think later (or not at all)? Can we, should we alter student perceptions about what math is and why it's important? Can we really teach problem solving, when we are freed from the presumed burden of teaching routine computation skills? Does it matter what or how we teach, or is learning going on in spite of us?

Technology. How do CAS's fit with other computer support for instruction? What is the role of the CAS as a "tool to think with?" More generally, what is the role of experimentation and discovery in learning and doing math? Can CAS's be combined with other tools? What can we expect in future hardware and software, and at what price? Will this technology eventually be available to all, or will its existence accentuate differences among schools, among students within schools, among socio-economic groups?

Here are some partial and tentative answers (and a few more questions):

Teaching. Most of the CAS effort we see today centers around calculus, but only because calculus is the capstone of secondary and the cornerstone of tertiary mathematics. In principle, all of our algorithmic manipulations can be performed by machines—and that accounts for almost all of what is taught and tested, right up to the graduate level. The obvious implication is that we must teach much less of how to do the manipulations and much more of what the manipulations are for and how to use them. Can our existing faculties do that? Not without major adjustments in their thinking about the subject and in their ways of relating to students. And that can't happen with faculty fully committed to teaching the "old" curricula full time.

Learning. We actually know very little about how people learn mathematical concepts and their uses. We think of our own experiences as models, but most of us are "exceptions" to the usual student experiences. We probably would have learned mathematics no matter what was going on in the classroom—and maybe we did. Was it important to learn how to compute square roots by hand? I recall taking some pleasure in the fact that I could do that when most of my classmates couldn't, but I can't do it now, and I don't mourn the loss of the skill.

Students will treat as important those things we consider important enough to reward them for. It's our job to reinvent the system of rewards. There will always be some students, like us, who will become mathematicians in spite of bad teaching or because of encouragement from a good teacher. But there won't be enough unless we can find ways to encourage many more than we are reaching now. As *Everybody Counts* points out, this is the only country in the world where people believe (a) it takes special talent to understand quantitative concepts, and (b) it is socially acceptable to declare oneself a quantitative illiterate.

Technology. At present, CAS's don't fit very well with other important forms of educational computer use—witness the Schoenfeld classification. Computer support for old ways of thinking about mathematics (e.g., drill-and-practice) must wither and die. Most of the other concepts described by Schoenfeld must be merged with CAS's into a much better hardware/software platform for doing and thinking mathematics—an ultimate declarative language for mathematics—a transcendent technology.

I am optimistic that we can solve the crucial curricular problems and that we can develop appropriate technology to support those solutions. However, given the economic and political structure of education in this country, I am not optimistic that the benefits of our solutions will be made available where they are needed most. Early in the next century, our society will be dependent on a workforce that must be scientifically and mathematically literate and that will no longer be dominated by relatively affluent white males. How can we see that our wonderful technology and our new technologically oriented curricula are made available to everyone, and not just to an elite few?

With a proper investment of time, effort, and resources, we can get the horse back in front of the cart. Can we also decide where we want to go?

References

1. Alvarado, F. L. and Ray, D. J. (1988). "Symbolically-assisted Numeric Computation in Education," *International Journal of Applied Engineering Education*, 4.

2. Hosack, J. (1988). "Computer Algebra Systems," in [8].

3. National Research Council (1989). *Everybody Counts: A Report to the Nation on the Future of Mathematics Education*, National Academy Press.

4. Page, W. (1990). "Computer Algebra Systems: Issues and Inquiries," *Computers and Mathematics with Applications*, 19, 51–69.

5. Porter, G. J. (1988). "Preface: The Use of Computers in Mathematics Instruction: Past History Future Prospects," in [8].

6. Schoenfeld, A. H. (1988). "Uses of Computers in Mathematics Instruction," in [8].

7. Schoenfeld, A. H. (1988). "When Good Teaching Leads to Bad Results: The Disasters of 'Well-Taught' Mathematics Courses," *Educational Psychologist* 23, 145–166.

8. Smith, D. A., Porter G. J., Leinbach, L. C., and Wenger, R. H. (Eds.) (1988). *Computers and Mathematics: The Use of Computers in Undergraduate Instruction*, MAA Notes Number 9, Mathematical Association of America.

9. White, J. E. (1988). "Teaching with CAL: A Mathematics Teaching and Learning Environment," *The College Math. Journal*, 19, 424–443.

On Calculus and Computers: Thoughts About Technologically Based Calculus Curricula That Might Make Sense *

Alan H. Schoenfeld
University of California at Berkeley

1 Introduction

The past few years have seen an exponential growth in the time and energy mathematics faculty have devoted to issues of technology and instruction, both in general and in the calculus in particular. Indeed, the papers in this volume demonstrate a mathematical and technical scope that would have been hard to imagine half a dozen years ago, much less in the early 1970's, when I first used computers for mathematics instruction! I take the opportunity here to look back at such attempts. My goal in this article is not to focus on any particular approach or any particular software, but rather to take a broad view—to situate in a broad context the various computational systems and approaches seen in calculus instruction across the country. In other words, I intend to sketch out the forest; most of the papers in this volume will paint the trees. This brief introduction to calculus and computers has three main parts which, if nothing else, have a rather attractive expository symmetry:

Part I: On Calculus, with or without Technology

Part II: On Technology in Calculus

Part III: On Technology, with or without Calculus.

That three-part core is bracketed by an introduction, which you are now reading, and a brief set of

concluding comments regarding the character of future developments. By way of introduction, I elaborate a bit on my opening sentence regarding the recent calculus-and-computers explosion. The curricular ferment of the past half dozen years has multiple causes. Among them are the perceived crisis in calculus instruction itself; one needs only point to *Toward a Lean and Lively Calculus* [1] and *Calculus for A New Century* [10], as indicators of the wide-spread sentiment that something needs to be done about calculus. And equally important, there is the immediate technological catalyst for action: the fact that there now exist powerful computer-based systems that can do with ease much if not all of the symbolic manipulation content of the calculus sequence.

Approximately a decade ago, Wilf [12] wrote as follows.

> A new program has recently been made available for my little (personal) computer, one of whose talents seem worthy of comment here because it knows calculus; in fact, as you read these words some of your students may be doing their homework with it.

The program Wilf described was MuMath, which was capable of performing the symbolic manipulations that occupy most of our students for the bulk of their calculus instruction. Like many programs that have followed it, MuMath suffered from an unfriendly user interface. Putting such difficulties aside, Wilf predicted that before long more powerful programs would be available on widely accessible machines. He then raised some serious questions:

---

*This paper was partially supported by the U.S. National Science Foundation through NSF grants MDR-8550332, BNS-8711342, and USE-8953974. NSF support does not necessarily imply the Foundation's endorsement of the ideas or opinions expressed herein. The content of this paper was first presented at the St. Olaf Conference on calculus and computers, October 20–22, 1989. A modified version of the paper with the same title appeared in the conference Proceedings, [3].

Will we allow students to bring (such machines and programs) into exams? Use them to do homework? How will the content of calculus courses be affected? Will we take the advice that we have been dispensing to teachers in the primary grades: that we should teach more of concepts and less of mechanics? What happens when $29.95 pocket computers can do all of the above and solve standard forms of differential equations, do multiple integrals, vector analysis, and what-have-you?

Wilf was off by an order of magnitude on his price predictions, but his predictions about the available technology were very much on the mark. Programs such as *Mathematica*, *Maple*, MacMath, Milo, Macsyma, and SMP live up to the computational promises made less than a decade ago. They solve standard forms of differential equations, do multiple integrals, vector analysis, perform matrix algebra and graph complex functions for starters. These programs are serious tools for the research mathematician, as well as possible instructional aids for students. They have the potential to revolutionize mathematics instruction, as well as to change the nature of research in mathematics. Yet, we do not have a clear sense of how to use these tools effectively in the classroom. Save for a few efforts to employ technology in mathematics instruction (see, e.g., [9]) the issues raised by Wilf have gone largely unaddressed. But there is now a flurry of activity. A fair number of colleges and universities have begun experimenting with computer-based (or assisted, or enhanced) labs and curricula. NSF support has enabled some fairly large (multi-year, multi-person, and sometimes multi-institution) projects to get off the ground.

To sum things up in brief: we have the need, and the tools, to do something different. So, what do we do? Put simply, it's too early to know. To paraphrase an old saying, anyone who claims to have "the solution" to issues of calculus and technology is either deluded or lying. There are many promising directions, and without doubt many garden paths among them. This is a time for creativity and divergence, for a good deal of sharing and compare-and-contrast. It may also be a time, amidst chaos, when an unbiased and slightly distant view may help. So here goes.

2   Part I: On Calculus, with or without Technology

The point of this brief section is to remind us that mathematics and calculus instruction existed before computers came along, and that some of the main issues regarding "what to do in calculus" are invariant across the technologies we might teach it with. Indeed, to broaden the scope of discussion, I note that most of the issues are invariant across all undergraduate instruction—and specific to the calculus only in that the specific concepts we want our students to come to understand come from that particular domain.

Here, in broad terms, is why the "big picture" is important. I take it as a given that readers of this volume are familiar with the current state of mathematics education nationwide, as documented in *Everybody Counts* [5]. Since, however, much of the focus in the current "crisis" has been on K–12 mathematics, readers may have missed a crucial and perhaps, at first blush, surprising statement:

> Reform of undergraduate mathematics is the key to revitalizing mathematics education.

On reflection it's easy to see why the statement is true—and it's easy to see the pivotal role that calculus reform plays in reshaping the entire undergraduate curriculum. Here are two of the main reasons calculus reform is so important.

First, despite the inroads made by courses in discrete mathematics and other options now available at the lower division level, calculus remains the main gateway to college mathematics. Students' experience in calculus often shapes the mathematical directions their careers will take. In the 1960's, slightly less than 5% of those entering college declared an intention to major in the mathematical sciences. Today the corresponding figure is less than 1%, which is roughly the percentage of students who graduate with majors in the mathematical sciences. There is, as we know, a large amount of turnover. We lose many of our intended majors because of our calculus instruction; we gain the majors we do because they have found something attractive about mathematics in the courses they take. A more interesting, meaningful, and ultimately mathematical calculus course (instead of the mostly mechanical ones that predominate today) could serve as the best way to keep potential majors interested in mathematics, and to lure new ones.

Second, calculus is the last mathematics course that most elementary and middle school teachers take before they begin their teaching careers. In other words, our calculus courses may be our last chance to show the next generation of teachers what mathematics is all about—and to provide them with models for mathematics instruction. If prospective teachers experience mathematics as a dead and deadening discipline, they are likely to pass the same kind of experience (as well as their dislike of the subject) on to their students. If, on the other hand, they experience mathematics as something lively and interesting— something they can make sense of—then they just might teach in a manner consistent with that stance. So in a very serious way, the fate of mathematics instruction at the beginning of the pipeline also rests in our hands.

Given this context, I quote the Goals for Instruction from the MAA Notes volume, *A Source Book for Teaching College Mathematics* (Schoenfeld [6]).

Goals For Instruction

1. Mathematics instruction should provide students with a sense of the discipline—a sense of its scope, power, uses, and history. It should give them a sense of what mathematics is and how it is done, at a level appropriate for the students to experience and understand. As a result of their instructional experiences, students should learn to value mathematics and to feel confident in their ability to do mathematics.

2. Mathematics instruction should develop students' understanding of important concepts in the appropriate core content. Instruction should be aimed at conceptual understanding rather than at mere mechanical skills, and at developing in students the ability to apply the subject matter they have studied with flexibility and resourcefulness.

3. Mathematics instruction should provide students the opportunity to explore a broad range of problems and problem situations, ranging from exercises to open-ended problems and exploratory situations. It should provide students with a broad range of approaches and techniques (ranging from the straightforward application of the appropriate algorithmic methods to the use of approximation methods, various modeling techniques, and the use of heuristic problem solving

strategies) for dealing with such problems.

4. Mathematics instruction should help students to develop what might be called a "mathematical point of view"—a predilection to analyze and understand, to perceive structure and structural relationships, to see how things fit together. (Note that those connections may be either pure or applied.) It should help students develop their analytical skills, and the ability to reason in extended chains of argument.

5. Mathematics instruction should help students to develop precision in both written and oral presentation. It should help students learn to present their analyses in clear and coherent arguments reflecting the mathematical style and sophistication appropriate to their mathematical levels. Students should learn to communicate with us and with each other, using the language of mathematics. For a more extensive discussion of these issues see Sterrett [11].

6. Mathematics instruction should help students develop the ability to read and use text and other mathematical materials. It should prepare students to become, as much as possible, independent learners, interpreters, and users of mathematics.

Now, everything in that list of goals applies to calculus; all that's missing is the specification of what we want to teach. When we turn our attention to the specific content of the calculus course, I have two main comments. The first is about the curriculum in general. I reiterate a theme strongly emphasized in the *Lean and Lively Calculus*: Any calculus reform must result in our instruction focusing carefully on a few fundamental ideas, instead of giving a large number of topics rather superficial coverage. The second is specific to technology and will be pursued below. Thanks to the technology, we can now have students engaged with more, and different, mathematics than they could have encountered in the pre-computer era. Hence decisions about what to teach, as well as how to teach it, will be affected by extant technologies.

3 Part II: On Technology in Calculus

It seems to me that once you know what you want to teach—a decision that may, as noted above, be affected by the presence of various technologies—then the primary question to ask about the use of technology in instruction is:

> How and where can we use technology to help us achieve the goals listed above, in ways that are better than we could do without it?

These issues are not new. Let me give two illustrations that date back to 1973, when I was teaching calculus at The University of California, Davis. My goal is to illustrate ways that technology can serve as a means of bootstrapping into more detailed investigations of the mathematical concepts we want our students to learn.

My first example comes from the first semester of calculus. The topic was approximations. I began class by deriving the formula for approximations using the first derivative,

$$f(x_0 + h) \sim f(x_0) + hf'(x_0).$$

Then we worked through a few of the standard examples, e.g., computing the square root of 63:

$$f(x) = x^{1/2}; \quad x_0 = 64; \quad h = -1,$$

$$f'(x) = (1/2)x^{-1/2}, \text{ so}$$

$$f(63) \sim f(64) + (-1)f'(64) = 8 - (1/2)(1/8) = 7.9375.$$

As it happens, this is a pretty good approximation. A better approximation, value, which I can obtain with a few button clicks of the calculator desk accessory on my Macintosh as I type this paper, is 7.9372539331838. So far, so good.

The class did a few more examples, Note that most of the approximation examples you find in texts, like the one given above, are artificial. That is, if you wanted to get the answer, you could use simple alternative methods to provide it. (Certainly you can find the square root of any number, to arbitrary accuracy, without difficulty. Moreover, the particular method used here is useful for computing $f(63)$, but not very useful for $f(73)$; for the method to work well you need a convenient value of x_0, one which makes h small.) So, the class decided to try an example that wouldn't be so easy to compute. How about estimating $(1.03)^{100}$?

Again, the computations are straightforward. We set $f(x) = x^{100}$, choose $x_0 = 1$ and $h = .03$. With $f'(x) = 100x^{99}$, we obtained

$$(1.03)^{100} \sim 1^{100} + 100(.03)(1^{99}) = 1 + 3 = 4.$$

And now, the calculator says that $(1.03)^{100} = 19.218631980865$. Hmm.....

In my classes, this gave rise to an interesting dilemma: the result derived from theory and the result from the black box calculator don't seem to agree. How can we resolve the conflict? The answer, of course, is that the quality of the approximation using $f'(x)$ depends on how flat $f(x)$ is at x_0; if the higher order terms in the Taylor approximation to $f(x)$ at x_0 are large, the approximation may not be very good. In particular, the relevant expansion for the function at hand is

$$(1 + h)^{100} = 1 + 100h + (1/2)(100)(99)h^2 + (1/6)(100)(99)(98)h^3 + \cdots.$$

and the first half dozen higher-order terms make a significant contribution to the value of the function at $h = .03$.

In reflecting on this example, I'd point out there are various stances one could take toward the use of the calculator in this kind of instructional context. One might ban it, saying that the students are supposed to learn the approximation formula and that the calculator has no place in the class. Or, one might decide to abandon the unit on approximations, since the calculator provides better values than the formula does, thus making the unit superfluous. Both of these strike me as extreme and wrong-headed. The role the calculator served in the class session described above was to point out what mileage you could get from a particular mathematical technique, and then to point out its limitations. As such, it helped me and my students get more deeply into the subject than we might otherwise have gone.

My second example, also dating back to 1973, also uses series—but this time a computer instead of a calculator for the detailed mathematical investigation. It concerns a standard example regarding the behavior of a power series at a distance from the point around which the series is expanded. Consider the Maclaurin series for $\sin x$,

$$\sin(x) \sim x - x^3/6 + x^5/120 - x^7/5040 + \cdots$$

How well does the approximation work? For $x = .1$, it works quite well. The first three terms alone produce

$$\sin(.1) \sim .1 - .0001666666 + .0000000833 - \cdots,$$

which is approximately .0998334833, versus the actual value (to 15 places) of .099833416646828 produced by the calculator. All well and good. But now try $\sin(\pi)$.

The first four terms yield

$$\sin(\pi) \sim \pi - \pi^3/6 + \pi^5/120 - \pi^7/5040$$
$$\sim 3.1415925 - 5.1677127 + 2.550164 - .5992645$$
$$=?????$$

Of course, we know why the series hardly looks like it's converging after the first four terms. In general the Taylor polynomials make for good approximations near the point around which the series is expanded, but they can diverge pretty wildly as you move away from that point. Now, in the days of paper-and-pencil or simple calculators, my classroom lessons could get about as far as I've described: We could get a sense that something was wrong, but not too much of an idea of why. Of course, I could draw the polynomial approximations to $\sin(x)$ on the board, or even show the students graphs of the Taylor approximations. But that didn't seem to help much: in the former case my sketches were too inaccurate to be compelling, and in the latter case, when everything is revealed in a single picture, there may be too much for students to make sense of. But even in 1973, I could have my students go to a terminal and graph the Taylor approximations to various functions, and see how they converged—over what intervals, and how fast. This hands on approach made the topic more accessible and meaningful. It also provided serious motivation for characterizing the error terms and for the concept of uniform convergence. As such, it may have made new stuff—stuff formerly considered too advanced, or too hard—accessible to students in an elementary course.

Having presented these two (rather old and primitive, by today's standards) examples, let me move up a level in abstraction and consider the kinds of changes that technology might bring to our calculus classes. There is a continuum of potential change. I characterize three kinds of change, in degree of increasing radicalism.

First, one might simply exploit extant technologies to do the traditional calculus better. My first, relatively trivial example of calculator use was of that sort. Here we face, at some level, the kinds of issues that school teachers faced when calculators became available. In the K–12 curriculum, problems and problem contexts always had to be simple, since students had to be able to work with the numbers (and symbols) at hand. Once calculators became available, that constraint went (more accurately, should have gone!) out the window. Using the technology,

students can, for example, find the statistical parameters of large sets of numbers without nice properties as easily as they could previously have found the statistical parameters of a much smaller set of cooked data using paper and pencil. Our choice of problems no longer needs to be constrained by the fact that the numbers must be simple, so students can arrive at answers in short order with pencil and paper. Now we can motivate much more interesting questions, and pursue them at much greater length. When your symbolic manipulation package can invert a 17×17 matrix without batting an eyelash, or graph an incredibly complex symbolic expression and allow you to determine its critical values empirically, or provide numerical or closed-form integrals, you can certainly engage the subject matter in much more substantial ways.

Second, one can reorganize and redesign the traditional calculus course, taking advantage of the technology to change the very nature of calculus instruction. Lynn Steen put the issue succinctly

> Should calculus be a laboratory course? Computer graphics now make possible the visual presentation of many of the dynamic phenomena studied in calculus. This unprecedented capability suggests wonderful pedagogical possibilities. Can we afford to provide every calculus student with access to a powerful workstation for calculus learning? Can we afford not to? [1, p. 160].

You need not look far to see a range of laboratory courses. Here are a few examples, all of which were discussed in the at the St. Olaf Conference. Ed Dubinsky uses ISETL to allow students to define and operate directly on mathematical functions (see Dubinsky's paper in this volume). Doug Child has students generate sequences that are computed and plotted both as $(n, A(n))$ pairs and plotted on one axis. Carl Leinbach uses a lab to emphasize the meaning of the fundamental theorem of calculus (Leinbach's paper is included in this volume). Joan Hundhausen explores the numerical solutions of differential equations. With the help of a computer algebra system, Ian McGee's freshmen in a calculus course tackle a problem whose solution requires facility with concepts such as volume of a solid of revolution, polynomial curve fitting, and function minimization. D. Brown, H. Porta, and J. J. Uhl have students compare the graphs of $[\sin(x + .01) - \sin(x)]/(.01)$ and $\cos x$, and draw the relevant conclusions. And so on. In short, the old calculus can be revitalized as a "hands on" activity. [See Part III for caveats.]

Third, and most extreme, one can redefine the calculus. As one example, Ken Hoffman and his colleagues at Hampshire College are members of an NSF-supported consortium of schools devoted to defining what might be called a calculus of (rather than for) the late 20th century. Their approach is more or less as follows. Rather than taking the content of the traditional calculus course and asking how you might do it better, begin with only a commitment to the fundamental mathematical ideas of calculus—differentiation and integration for starters. Next, go out into the real world, and see how those fundamental ideas are being implemented in practice. And then create a course that reflects today's calculus, not Newton's. Such a course may be heavily applied, with quite large problems worked by teams of students—since that's the way the ideas of calculus are used outside of classrooms.

4 Part III: On Technology, with or without Calculus

I have no particular axe to grind with regard to the use of technology in calculus courses. On the one hand, I obviously favor the appropriate uses of technology. As noted above, I explored the uses of calculators and computers in calculus courses in the early 1970's; I argued above, with conviction, that there is unquestionably a major role for technology in today's instruction. In fact, my research group has developed some fairly sophisticated software (Schoenfeld [7]) to help students learn about mathematical functions and graphs. On the other hand, I am hardly a naive technophile or technology booster.

For the past few years my research group has been taking a close look at what actually happens when students sit down and work with state-of-the-art technology (Schoenfeld, Arcavi, and Smith [8]). The reality of student-computer interactions is somewhat chastening. To put things simply, a number of naive assumptions about the wonderful things technology can do for students just don't stand up to close scrutiny. In this section of the paper, I offer some examples to indicate some of the issues that will have to be dealt with regarding the use of technology in instruction. We need to keep such examples in mind, for the presence of technology itself will not suffice to provide students with the kinds of learning experiences they need.

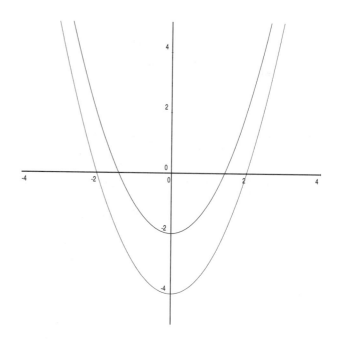

Figure 1 : Vertically translated functions that don't look it.

Here are a few of the issues that emerged in our research.

Issue 1: (Near-)Perfect representations of graphs on the computer screen may give students the wrong impressions! We all know about the limitations of hand-drawn graphs on paper or chalkboard. No matter how accurately we try to sketch a function $f(x)$ and the function $g(x) = f(x) + C$, it takes an act of faith to believe that $g(x)$ is truly a vertical translation of $f(x)$—in the sense that you could pick up the graph of $f(x)$ as a rigid body, and move it C units vertically to produce the graph of $g(x)$. One might suppose that the very accurate graphs produced by computers would convince students on this point; after all, their accuracy is one of the most highly touted virtues of computers. But one would be wrong. Figure 1 shows two functions that are indeed vertical translations of each other. Alas, they don't look it! The human perceptual apparatus tends to measure "set distance" (the minimum distance from a point $f(x_0)$ to the curve $g(x)$) rather than vertical distance (the distance from $f(x_0)$ to the point $g(x_0)$), so $f(x)$ and $g(x)$ appear closer at the boundaries of Figure 1 than they do near $x = 0$. It takes "measuring bars" as in Figure 2 to help overcome the illusion.

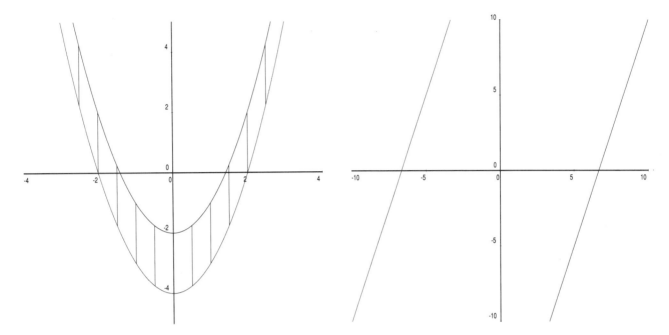

Figure 2 : Vertical measuring bars help
rectify the optical illusion.

Figure 3 : Vertical translates appear to move
in near-horizontal fashion.

Here is a second illustration of the same theme, discussed at some length in Goldenberg *et. al.* [2]. Figure 3 illustrates the graphs of the functions $3x - 20$ and $3x + 20$ on a coordinate grid extending from $x = -10$ to $x = +10$. Goldenberg's intention was for students to see such graphs as vertical translates of each other, with their positions determined by their y-intercepts. Instead, however, students perceived the functions to move more along the horizontal than the vertical axis!

In short: We can't assume that students will see what we want them to see, even if it's accurately represented on the screen. The pedagogical moral to this story is obvious: The job of the teacher remains that of helping students to see the mathematics as we understand it. The presence of the technology may make it easier for students to see some things and may actually be misleading in other cases. In all cases, it's our job to structure the instructional interactions so that students make sense of what they see in the appropriate ways.

The software my research group developed allows students to vary the parameters in symbolic expressions of functions, and to watch the graphs change in real time as they do. As the simplest example of what the system can do: you can change the m or b in the expression $y = mx + b$, and watch the line move up and

down or rotate around $(0, b)$ respectively. For those of us, who know what to look for, the correlation between the symbolic and graphical actions makes sense. Students may miss it altogether, because they don't know what to focus on when they look at the screen!

Issue 2: Just because something is up on the screen doesn't mean students see it or make sense of it.

Here's a slightly more complex example. The function $f(x) = ax^2 + bx + c$ can also be expressed in the form $f(x) = a(x - b')^2 + c'$ and, if f has real roots, as $f(x) = a(x - b'')(x - c'')$. The first is easy to work with algebraically, but yields little direct information about properties of its graph (save that the sign and magnitude of a determine the width and direction of the parabola). The second provides the additional information that the vertex of the parabola is at the point (b', c'), and the third that the graph of $f(x)$ crosses the x-axis at the points $(b'', 0)$ and $(c'', 0)$. In our software all three algebraic forms, as well as the graph, are simultaneously displayed; as students change the parameters in any of the algebraic forms, the graph changes and the other algebraic forms are updated automatically. For the mathematician, this dramatically makes the points that (a) the different algebraic forms all represent the same function, and (b) those forms provide different information about the function. To put things simply, mathematically naive students missed the point. There was too much going on up the screen, and they ignored (or failed to see) what they didn't understand.

Issue 3: Don't expect students to make sense of stand-alone or self-contained micro-worlds all by themselves—even when the students are ready for them.

The one-liner here is that it's easy to underestimate the amount of structure or guidance that students need in order to "get into" an environment. Remember how unsatisfying some of your biology, chemistry, or physics labs were when you were either (a) underprepared, or (b) able to go through the motions without really being engaged in the subject matter? It's no different in high-tech mathematics labs.

Issue 4: Be careful in making assumptions about the nature of the objects the student is working with, or the properties they attribute to them.

It's difficult to tell this story in brief; see Schoenfeld, Arcavi, and Smith [8] for details. Here are a few of the mistaken impressions students of various ages develop when working with computer-based graphing systems. First, when a graphing system draws a line dynamically (as opposed to having it simply appear on the screen), some students think that the line "has a direction" and "comes from" some place off screen. A common misapprehension is that lines of positive slope "come from" the bottom of the screen, lines of negative slope from the top (since lines are often drawn left-to-right). Similarly, naive students frequently focus their attention on the boundaries of the screen (where lines come from) rather than the axes; hence discovery exercises (e.g. "graph the lines $y = 2x$, $y = 3x$, $y = 4x$, etc., and look for a pattern.") may be much tougher than one would think.

I could go on at length, but these examples should suffice to make the point. The presence of computer-based technologies for doing mathematics will undoubtedly make some aspects of instruction much easier. However, naive assumptions about magic happening as a result of putting the technology in the hands of the students are unwarranted. The bottom line is this: Whether we're talking about calculus instruction with or without the help of technology, the underlying issues remain the same. What counts is what students experience, and what sense they make of it. Though show-and-tell may be much more dra-

matic with new tools, and students may be able to crank out solutions to much more complicated problems with them, we still have to figure out just what it is that we want the students to learn, and what set of experiences (with the technology properly exploited) will give them the best chances at learning it. This conceptual task will not be any easier for the presence of the technology—but the results, given the potential of the new tools, may be far more exciting and more enlightening for our students.

5   Concluding Comments

One recent attempt to come to grips with change in the calculus curriculum was an NSF-supported conference entitled "Calculus and Computers: Toward a Curriculum for the 1990's" held at Berkeley from August 24 though August 27, 1989. Participants at that conference considered a broad range of issues related to technology and calculus instruction. There, as at a dozen or more similar conferences over the past few years, people presented background papers, gave demonstrations and workshops, and discussed the features of their innovative and often technology-based calculus courses. In addition, the concluding sessions featured group discussions in which the participants looked for a consensus about "what works," and attempt to define an R&D agenda for making progress on the issues of calculus and computers. The spirit of that conference, and its closing recommendations, serve both as an appropriate close to this paper and as an overture to others in this volume.

Conference participants recognized that this is a time of flux and exploration. "Over the next few years we hope to come to understand much of the technology's mathematical and pedagogical potential. Right now, however, we believe that the most sensible approach is broad-based, experimental, and incremental" (Linn, Ribet, and Schoenfeld [4], p. 7). That is, there is no single right answer available at present, or even a small number of approximate solutions. Hence a range of ideas and approaches should be nurtured and encouraged to flourish. Second, this must be a time of open exchange and communication. There is no sense in having independent groups reinvent the (sometimes square or oblong) wheel, and information exchange may be critical to the success of the enterprise. It is in the best interests of educational researchers, those developing innovative curriculum materials, and those interested in improving calculus instruction at a range of institutions to share ideas from the very beginning. Volumes such as this work toward those goals. Turn the pages and see.

References

1. Douglas, R. G. (ed.) (1986). *Toward A Lean and Lively Calculus: Report of the Conference/Workshop to Develop Curriculum and Teaching Materials for Calculus at the College Level*, MAA Notes Number 6, Mathematical Association of America.

2. Goldenberg, P. (1988). "Mathematics, metaphors, and human factors: Mathematical, technical, and pedagogical challenges in the educational use of graphical representation of functions," *Journal of Mathematical Behavior*, 7, 135–173.

3. "Proceedings of the St. Olaf Conference." (1989). (12 papers on Symbolic Computing and Undergraduate Mathematics), c/o Paul Humke, Department of Mathematics, St. Olaf College.

4. Linn, M. C., Ribet, K. A., and Schoenfeld, A. H. (1989). *Calculus and computers: Toward a curriculum for the 1990's*, Final report of NSF grant USE-8953974.

5. National Research Council. (1989). *Everybody Counts: A Report to the Nation on the Future of Mathematics Education*, Washington, DC: National Academy Press.

6. Schoenfeld, A. H. (ed.) (1990). *A Source Book for College Mathematics Teaching*, Mathematical Association of America.

7. Schoenfeld, A. H. (in press). "GRAPHER: A Case Study in Educational Technology, Research, and Development," in A. diSessa, M. Gardner, J. Greeno, F. Reif, A. Schoenfeld, and E. Stage (eds.), *Toward a Scientific Practice of Science Education*, Erlbaum, Hillsdale, NJ.

8. Schoenfeld, A. H., Arcavi, A. A., and Smith, J. P. (in press). "Learning," to appear in R. Glaser (ed.), *Advances in Instructional Psychology*, Vol. 4, Erlbaum, Hillsdale, NJ.

9. Smith, D. A., Porter, G. A., Leinbach, L. C., and Wenger, R. H. (eds.). (1988). *Computers and Mathematics: The Use of Computers in Undergraduate Instruction*, MAA Notes Number 9, Mathematical Association of America.

10. Steen, L. A. (ed.) (1988). *Calculus for A New Century*, MAA Notes Number 8, Mathematical Association of America.

11. Sterrett, A. (ed.) (1990). *Using Writing to Teach Mathematics*, MAA Notes Number 16, Mathematical Association of America.

12. Wilf, H. (1982). "The Disk with the College Education," *American Mathematical Monthly*, 89, 1, 4–8.

Symbolic Computing in Undergraduate Mathematics: Symbols, Pictures, Numbers, and Insight *

Paul Zorn
St. Olaf College

1 Introduction

"The purpose of computing is insight, not numbers."

Richard Hamming's aphorism reminds us that although computers easily—all too easily—produce floods of numbers, the numbers themselves are not really the point. What the numbers **say**, if we can figure it out, makes computing worth doing.

Hamming's motto is partly a warning: don't get lost in the numbers. But it's also a promise: somewhere in the ore of raw numbers, nuggets of insight are waiting to be found. I want to argue here that with symbolic computing systems, we can extend and strengthen Hamming's point. Symbolic computer systems, such as *Mathematica*, generate not only numbers, but also pictures and symbols. Numerical computing offers numerical insights. Symbolic mathematical computing offers something new and stronger: insights informed by the **combination** of numbers, pictures, and symbols.

Often we treat mathematics students like fresco painters' apprentices in 15th century Florence: Only after years of the mathematical equivalent of brush-cleaning and plaster-mixing (and perhaps an occasional inconspicuous cherub) do we let them in on the big picture. This analogy is admittedly dangerous; after all, Florence somehow did very well for itself. But even if this system worked in the fresco business, it doesn't seem to in mathematics. Hardly anyone will hang around that long. Letting our apprentices in on the big picture a bit earlier might encourage more of them to stay.

---

*This paper is revised from the text of an hour address given at the Joint MAA/AMS Summer Meetings at Columbus, August 1990.

2 First Examples: Visualizing Complex Analytic Functions

Elementary complex analysis is a subject at once beautiful and difficult to envision. For those very reasons, high-level computing offers special advantages.

A complex analytic function is a **pair** of real-valued functions of two variables, linked in a special way. The Cauchy-Riemann equations tell how: Let

$$f(z) = f(x + iy) = u(x, y) + iv(x, y)$$

be a complex-valued function of a complex variable, with real and imaginary parts u and v. A basic (probably **the** basic) fact of complex analysis is that f is analytic only if u and v satisfy the Cauchy-Riemann equations:

$$\frac{\partial u}{\partial x} = \frac{\partial v}{\partial y}, \qquad \frac{\partial u}{\partial y} = -\frac{\partial v}{\partial x}.$$

Put this way, the result is not very suggestive. Put geometrically, things are much more interesting. The Cauchy-Riemann equations say (among other things) that f is analytic in a domain only if the gradients ∇u and ∇v are perpendicular at each point $z = (x, y)$. Two *Mathematica* graphs show nicely what this means for a typical analytic function, $f(z) = \sin z$. First, the definitions:

```
In[1]:= f[z_] := Sin[z]
In[2]:= z = x+Iy; u = Re[f[z]];
        v = Im[f[z]];
```

Now we plot the surfaces corresponding to u and v:

```
In[3]:= Plot3D[u,{x,-5,5},{y,-2,2}]
Out[3]=
```

17

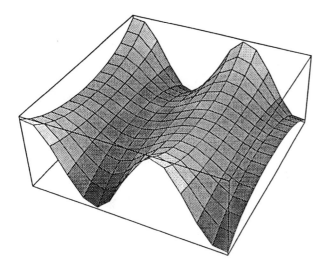

```
In[4]:=  Plot3D[v,{x,-5,5},{y,-2,2}]
Out[4]=
```

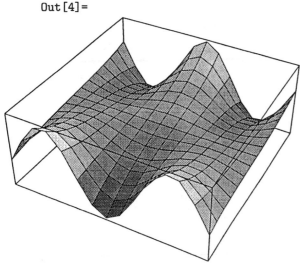

The pictures show that the gradients are indeed perpendicular at corresponding points: the "uphill" direction on one surface corresponds to a "level" direction on the other. With a bit more effort, other facets of the duality between the real and imaginary parts of a complex number reveal themselves. For example, the Cauchy-Riemann equations say more than that ∇u and ∇v are everywhere perpendicular. They say also that at each point z in the domain, ∇v is the result of **rotating ∇u 90° counterclockwise.** The surfaces, examined more closely, reveal this orientation.

The same pictures illustrate an important property of harmonic functions. It follows immediately from the Cauchy-Riemann equations that both the real and

imaginary parts of an analytic function are harmonic, i.e., satisfy Laplace's equation. A basic property of harmonic functions is the maximum-minimum principle: a non-constant harmonic function can attain neither a maximum nor a minimum value in the interior of its domain. Both of the surfaces above illustrate this: maximum and minimum values are attained only at the boundary of the domain. A contour plot, say of u, shows this even more clearly; all the level curves continue to the boundary of the domain:

```
In[5]:=ContourPlot[u,{x,-5,5},{y,-2,2}]
Out[5]=
```

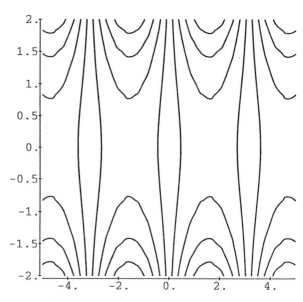

2.1 Conformal Mapping, Geometrically

Another excellent way to visualize complex functions is as mappings. Recall, first, an elementary property or two of conformal maps. First, every one-to-one analytic function **is** conformal—i.e., it preserves angles between curves. What does this mean, for example, about how a particular one-to-one analytic function, say

$$f(z) = \frac{i+z}{2-z}$$

transforms the square $[-1,1] \times [-1,1]$? Another *Mathematica*-generated picture will help; it shows the **image** under f of a 25×25 rectangular grid on $[-1,1] \times [-1,1]$:

```
In[6]:= CartesianMap[f,{-1,1},{-1,1},
          PlotPoints -> 25]
Out[6]=
```

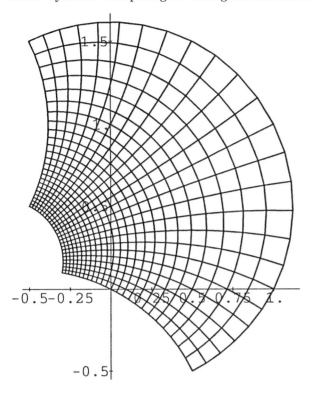

As conformality predicts, in the **image** of the square grid (as in the grid itself) all the angles are right angles. The image of a 25×25 polar grid has similar properties:

```
In[7]:= PolarMap[f,{0,1},{0,2Pi},
          PlotPoints -> 25]
Out[7]=
```

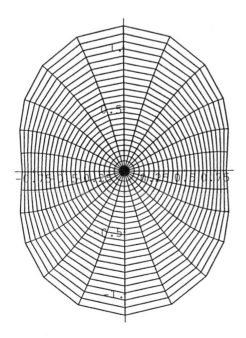

Taken together, the pictures offer a bonus insight: f seems to map all the **original** grid lines and circles to **new** lines and arcs of circles. Or so it appears—one can show, analytically, that this property holds for **every** complex map of the form

$$l(z) = \frac{az + b}{cz + d}.$$

Back, now, to words—dry as they seem after such pictures.

2.2 Symbolic Computing: Chemistry or Alchemy?

Extravagant, even metaphysical, claims have always been made for new technologies. During the industrial revolution, the mind was sometimes depicted as an exquisitely complicated clockwork device. Today, the idea of gears and pulleys in the head sounds laughably quaint; clearly, the brain is some sort of exotic computer.

Even allowing for due caution, it's hard to deny the significance of mathematical computing. Already, computing has profoundly changed the mathematics we do—**what** we do, **how** we do it, and even **why** we do it. How we **teach** mathematics has been a lot slower to come along. In a sense, that's not surprising; maybe it's even appropriate. Since modern computing—symbolic computing in particular—does so much of what we've traditionally taught, at least through the undergraduate level, the pedagogical questions that arise are extremely important, but extremely difficult. What should we teach? What should students know? What does it **mean**, anymore, to know mathematics? How many skills, and which ones, must students know? What is conceptual learning, how do we promote it, and how do we measure it?

One thing, at least, is clear: Like it or not, symbolic computing can't be ignored. Even if we do, our students won't. For good or evil, the power and flash of symbolic computing are here to stay. That flash could be gold or it could be baser stuff; whether we use it for chemistry or alchemy is up to us.

First, a piece of terminology. By "symbolic computing," I mean not just the explicit manipulation of symbols by computer, but rather all the sorts of things that a system like *Mathematica* does: computations in all three modes—graphic, numeric, and algebraic. The part—symbolic computing—stands for the whole—high-level mathematical computing in general.

My use of *Mathematica* is in the same spirit: as a canonical example of the full class of systems I have in mind. (This is not to endorse a product. *Maple*, *Derive*, MACSYMA, and other programs that combine graphic, symbolic, and numeric modes of computing could do equally well.)

2.3 This Paper: an Overview

First, some things this paper **isn't**:

- It isn't a comprehensive demonstration of symbolic computing.

- It isn't an overview of projects that use symbolic computing. I'll mention some, but only as examples. For more information on projects, see the references, especially [1], [5], and [6].

- It isn't a collection of practical advice. (For information on practical matters, see especially Z. Karian and A. Sterrett's paper in [6], and much of [5].)

The paper **is** about more general issues:

- What needs fixing?

- What, uniquely, does symbolic computing do? Why should we use it?

- What are the costs of symbolic computing? Why should we pay them? How should we pay them?

- How can we **assess** whether symbolic computing works for students?

In telegraphic form, here are my answers:

- We teach mathematical **algorithms**—symbolic algorithms on symbolic data, anyway— reasonably well. We teach mathematical **ideas** poorly.

- Since symbolic computing handles most of the standard algorithms of undergraduate mathematics, it should free time and effort for other things. We should use it both for horsepower— doing some of the same things faster and better— and, more importantly, to do different, mainly graphical and numerical, things.

- As with the savings and loan crisis, the costs are real; so are the costs of doing nothing. On budget or off budget, we'll pay. We **should** pay because symbolic computing has a real chance to foster better conceptual learning.

- Assessing our success will be difficult, time-consuming, and contentious. We'll need all the help we can get from educational researchers. We can help them by framing and asking the right questions.

The rest of this paper aims to support these answers. There will also be more flashy pictures.

3   Examples from Calculus

The following examples come from elementary calculus. Commands appear in a *Mathematica*-like syntax; some involve special-purpose commands, easily programmable in *Mathematica*. The examples are chosen both for their simplicity and to illustrate the power of combining graphic, numeric, and algebraic points of view on topics in calculus.

```
In[1]:= Integrate[Cos[x^2],x]
Out[1]= Integrate[Cos[x^2],x]
```

Nothing happens, but for a good reason: the integrand has no elementary antiderivative. Still, the integrand is perfectly tractable; the definite integral $\int_0^1 \cos(x^2)\,dx$ should make sense.

```
In[2]:= Graph[Cos[x^2],{x,0,1}]
Out[2]=
```

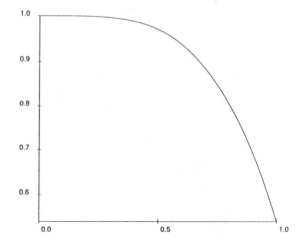

At a glance, the estimate $\int_0^1 \cos(x^2)\,dx \approx .9$ looks reasonable. Let's refine it:

```
In[3] := MidGraph[Cos[x^2],{x,0,1},10]
Out[3]=
```

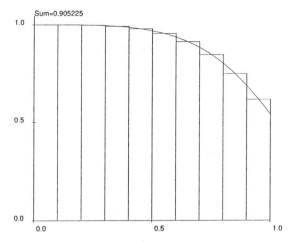

The midpoint estimate $\int_0^1 \cos(x^2)\,dx \approx .905225$ looks quite convincing. Judging from the graph's concavity, we have a (slight) overestimate. Let's try another tack:

```
In[4] := Taylor[Cos[x^2],x,0,4]
Out[4]= 1 - x^4/2

In[5] := Integrate[%,{x,0,1}]
Out[5]= 9/10
```

The fourth degree Taylor approximation to $\cos x^2$ is easy to compute, and easier still to integrate. But how good is the new estimate? A picture would help:

```
In[6] := Graph[ Cos[x^2],1-x^4/2,
            {x,0,1}]
Out[6]=
```

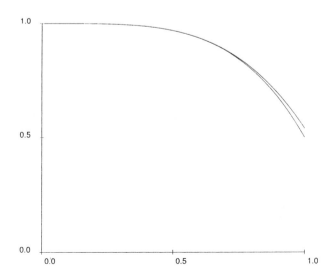

The **upper** graph is of $\cos x^2$. Thus the approximation

$$\int_0^1 \cos(x^2)\,dx \approx \int_0^1 \left(1 - \frac{x^4}{2}\right)\,dx = .9$$

is, again, a (slight!) overestimate.

We could easily continue, generating closer approximations, calculating error estimates, etc. Instead, let's try something completely different:

```
In[7] := MonteCarlo[ Cos[x^2],
            {x,0,1},200 ]
Out[7]=
```

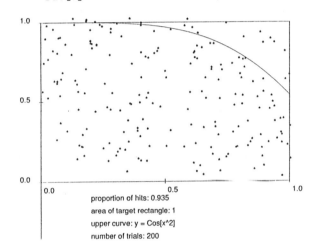

proportion of hits: 0.935
area of target rectangle: 1
upper curve: y = Cos[x^2]
number of trials: 200

Our probabilistic estimate, $\int_0^1 \cos(x^2)\,dx \approx .885$, is surprisingly good. The method is, granted, computationally extravagant; it required 200 function evaluations. But to illustrate, simply and convincingly, the connection between integral and probability, the computations are worth doing.

Using numerical techniques in calculus is not a new-fangled notion. In his 1922 edition of *Introduction to the Calculus*, Osgood treats numerical methods of solving equations at some length. After a leisurely (for that time!) introduction to the secant method, Osgood exhorts the student to remember the idea, not the formula, using a period version of the "desert island" argument:

> The method, once appreciated, can never be forgotten. If the student finds himself in a lumber camp with nothing but [Pierce's] tables at hand, he may solve his equation if he has once laid hold of the method. ... But this he may not do if he has relied on the formula in the book.

4   What's Wrong Now?

I'll use elementary calculus as a canonical example, but similar complaints could be made about other undergraduate courses.

Lots is wrong with the way we teach calculus. The worst problem, and the one most relevant here, is that we have reduced the subject (perhaps not as we **teach** it but certainly as students **learn** it) to a compendium of algebraic symbol-manipulation techniques. In effect, calculus becomes a species of algebra—not a very hardy species, at that.

Consider the notorious "methods of integration." The name itself bespeaks confusion: "antidifferentiation in closed form" would be more accurate. Students regard the fundamental theorem as a working **definition** of the definite integral—gratefully, given the pain the real definition causes. Their strategy is not bad, admittedly. Nearly every integral they see succumbs to closed-form methods.

That *Mathematica* antidifferentiates better than most of us is not, of course, any reason to jettison antidifferentiation methods wholesale. Antidifferentiation is good for more than computing integrals. Substitution and integration by parts, for example, are really instances of far-reaching mathematical principles. Even despised and rejected techniques, like the $\tan u/2$ substitution, can reveal useful properties of and relations among the elementary functions. Symbolic computing should not abolish antidifferentiation techniques, but put them in sensible perspective.

Antidifferentiation is important, but, all alone, it fails both to handle many integrals and, more important, to illustrate the integral adequately as an **analytic** object. With symbolic computing to support numerical and graphical points of view, a deeper, better balanced, and more applicable understanding of the integral might emerge.

The next example concerns the integral $\int x^{10}e^x\,dx$. This one can—but probably shouldn't—be done in closed form.

```
In[1]:= Integrate[ x^10 Exp[x], x ]
```

$$\text{Out[1]}= 3628800\ E^x - 3628800\ E^x\ x +$$
$$1814400\ E^x\ x^2 - 604800\ E^x\ x^3 +$$
$$151200\ E^x\ x^4 - 30240\ E^x\ x^5 +$$
$$5040\ E^x\ x^6 - 720\ E^x\ x^7 +$$
$$90\ E^x\ x^8 - 10\ E^x\ x^9 + E^x\ x^{10}$$

```
In[2]:= Integrate[ x^10 Exp[x],{x,0,1}]
Out[2]= -3628800 + 1334961 E
```

So far, so good; we found closed-form expressions for both the indefinite and definite integrals. Both contain large factorial-type constants. How large, numerically, is the definite integral?

```
In[3]:= N[%]
Out[3]= 0.228002
```

Should the small answer be surprising, given the colossal integers above? Not really: As the graph below shows, the integrand is small and quite tame.

```
In[4]:= Plot[x^10 Exp[x],{x,0,1}]
Out[4]=
```

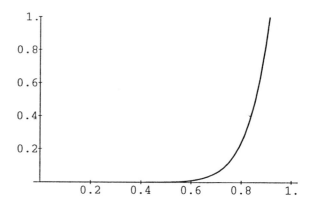

Integration by parts is, of course, the natural tool for handling both $\int_0^1 x^{10}e^x\,dx$ and $\int x^{10}e^x\,dx$. It leads to these "reduction" formulas:

$$\int_0^1 x^n e^x\,dx = e - n\int_0^1 x^{n-1}e^x\,dx$$

and

$$\int x^n e^x\,dx = x^n e^x - n\int x^{n-1}e^x\,dx.$$

Letting $i(n) = \int_0^1 x^n e^x\,dx$ and $j(n) = \int x^n e^x\,dx$, the recursion relations become:

$$i(n) = e - n\,i(n-1) \qquad j(n) = x^n e^x - n\,j(n-1)$$

The paperwork to compute $i(10)$ and $j(10)$ recursively is tedious. We'll leave most of it to *Mathematica*:

```
In[5]:= i[0] = Integrate[x^0 Exp[x],
        {x,0,1}]
Out[5]= -1 + E

In[6]:= i[n_] := E - n i[n-1]

In[7]:= i[10]
Out[7]= E - 10 (E - 9 (E - 8 (E - 7
        (E - 6 (E - 5 (E - 4 (-3
        (-2 + E) + E)))))))
```

The parentheses are confusing, but they arise honestly from the recursive computation. More transparently:

```
In[8]:= Simplify[%]
Out[8]= -3628800 + 1334961 E
```

We can compute the antiderivative, symbolically, in just the same spirit:

```
In[9]:= j[0] = Exp[x]

In[10]:= j[n_] := x^n Exp[x] - n j[n-1]

In[11]:= j[10]

          x 10        x 9
Out[11]= E  x   - 10 (E  x  -

            x 8       x 7
         9 (E  x  - 8 (E  x  -

            x 6       x 5
         7 (E  x  - 6 (E  x  -

            x 4       x 3
         5 (E  x  - 4 (E  x  -

            x 2       x
         3 (E  x  - 2 (-E  +
          x
         E  x)))))))))

In[12]:= Simplify[%]

          x
Out[12]= E  (3628800 - 3628800 x +

               2            3
         1814400 x  - 604800 x  +
```

```
            4            5
   151200 x  - 30240 x  +
          6        7        8
   5040 x  - 720 x  + 90 x  -
        9     10
   10 x  + x  )
```

Some symbol manipulations in calculus courses are, to be sure, far from trivial. Many require meticulous persistence; some call for real ingenuity. But the point remains: Calculus is, as a mathematical discipline, a branch of analysis, but most of what students **do**—limit computations, differentiation, integration, convergence testing, etc.—reduces to **algebra**—algebraic manipulation of algebraically presented functions. (The transcendental functions aren't algebraic, some will protest. This is a good point; unfortunately, few of our students know that.)

In the worst cases, even algebra drops away, leaving a curious sort of word-processing. Calculus becomes a collection of text-editing macros: replace "sin" by "cos", prepend a minus sign, etc. Not coincidentally, these are just the sorts of things machines do best.

4.1 What's Wrong with Symbol Manipulation?

As one student (quoted in one of those coffee-room "blooper" collections) charmingly put it:

> Algebraic symbols are what you use when you don't know what you're talking about.

That's about right, although not quite in the sense that the student meant. I'd put it differently:

> When they use symbols, students often don't know what they're talking about.

In other words, students don't really understand, in any useful or durable sense, what the symbols symbolize. If they did, we would never see "computations" like

$$\frac{\sin x}{x} = \sin.$$

Such computations are the predictable results of overstressing all-too-purely symbolic representations of mathematical ideas.

4.2 Good and Bad Representations of Functions: the Bounding Problem

The idea of function—so basic that we usually regard it as a pre-calculus subject—offers an important, painful case study in the dangers of symbol manipulation calculus. Students' understanding of functions—all functions—is surprisingly thin and full of gaps. Even those who adroitly differentiate, integrate, and l'Hospitalize fancy combinations of elementary functions are usually stumped by simple **non-algebraic** problems.

Bounding functions, although in some sense an "easy" idea, causes calculus students particular trouble. Given a function and an interval over which to bound it, weaker students use an "integer search" technique: plug in the endpoints and (if any come easily to mind) nearby integers. Better students use algebraic calculus to look for a maximum: set the derivative equal to zero, solve algebraically, etc. Even if everything works—by no means always the case—there may still be trouble sorting out critical points. If any hitch arises, most students are stumped.

Students find the bounding problem difficult for one main reason: it requires them to see the function explicitly as **acting on a set of inputs**, rather than in static, symbolic form. The symbolic representation of a function—even when a simple one exists—lends itself poorly to bounding problems. Representing functions symbolically is sometimes appropriate and efficient, as when performing formal operations, like differentiation. But symbolic representations are less suggestive, when, as in bounding problems, functions must be viewed dynamically, as processes, rather than statically, as objects.

To understand functions richly enough and flexibly enough to accommodate the various ways they arise in mathematics, students need to see (and manipulate, and use) functions as more than formulas. They should pass easily from one style of representation to another—symbolic, graphical, and numerical—manipulating functions as the situation requires—as objects, as processes, or as formal names.[1]

Reducing calculus to symbol manipulation distorts students' understanding of fundamental analytic ideas: rate and amount, derivative and integral, frequency and accumulation. For instance[2], more stu-

---

[1]See the work of Dubinsky. Dubinsky and others use symbolic computing, in disciplined ways, to help students "construct" rich mental images of functions and other mathematical objects. See, e.g., Dubinsky's paper in [6].

[2]I heard this example from Professor F. Tufte.

dents would probably evaluate

$$\int_{-1}^{2} x^2 \sin x \, dx$$

correctly than

$$\int_{-1}^{2} |x| \, dx.$$

Similarly, most students would translate the easy question of whether

$$\int_{1}^{\infty} \frac{1}{x^3 + 1} \, dx$$

converges into the hard problem of computing the limit algebraically.

5 How Symbolic Computing Can Help

Symbolic computing can act as a machete to help rescue the ideas of calculus (and other courses) from the strangling kudzu of symbol manipulations.

One way is simply by doing some of the manipulations students would otherwise do. Many of the calculations in mathematics courses and books are there, in theory, to illustrate or clarify ideas and principles. In fact, the computations often obscure, or even **become**, the point. (There is some irony here: Hamming's well-founded concern is that machine computing can obscure underlying ideas. In practice, hand computations are just as likely to get in the way.) A little mechanical help, used judiciously, might reveal more clearly the insights behind the numbers.

Another advantage of symbolic computing—the most important one, in my opinion—is the possibility of adding graphical and numerical viewpoints to the traditional symbolic one. The main ideas of calculus look quite different when viewed from various angles. If we mean seriously to teach students the concepts of calculus in usable forms, we should use all the tools we have to give these ideas depth and shading.

Numerical and graphical viewpoints are hard to pursue by hand, or even by book. Calculus books are, more and more, filled with stunning graphical and numerical displays. This is much better than nothing, but it is still only an evolved form of show and tell. With symbolic computing, graphical, numerical, and algebraic viewpoints are all available to students—not merely as demonstrations, but as tools. **They can not only observe, but also construct and**

manipulate mathematical objects graphically, numerically, and algebraically.

5.1 Infinite Series

The related themes of convergence and approximation are a case in point. In treating infinite series—an important "bridging" topic between elementary calculus and real analysis—standard calculus courses concentrate almost exclusively on the problem of convergence **testing**. Students may learn adroitly to affix the labels "convergent" or "divergent" to series without really understanding what **either** label means. In this never-never land, a series is said to converge if it passes one of the five or six standard tests, and to diverge if it fails. (There is always a niggling fear of series that pass one test and fail another, but as students quickly learn, this seldom happens.)

It seems natural to ask what a convergent series converges **to**, but the question usually gets short shrift. There are good reasons—approximation and error estimation are hard to treat without machine assistance—but the result is a rather ethereal treatment of the subject. As mathematicians, we enjoy such heady sensations, but many students experience them as free-floating anxiety and confusion.

Symbolic computing permits a more concrete, down-to-earth approach to series. With symbolic computing, we can take the problem of estimating limits of series, and defending those estimates, just as seriously as the abstract question of convergence. Such an approach is probably more useful for applications; more important, it should promote a deeper understanding of the analytic phenomenon of convergence. With, say, *Mathematica*, we could handle problems like these:

Example 1. It can be shown that $\sum_{j=1}^{\infty} \frac{1}{j^2} = \frac{\pi^2}{6}$. A formal proof is subtle, but numerical experiments are easy. Try some. Then use integration to find an n for which the nth partial sum S_n satisfies $|S_n - \pi^2/6| < .0001$.

Solution. First we define the series and its partial sums:

```
In[1]:= a[k_] := 1/k^2 ;
        s[n_] := Sum[a[k],{k,1,n}]

In[2]:= {s[10],s[100],s[200],N[Pi^2/6]}
Out[2]= {1.54977,1.63498,1.63995,1.64493}
```

The numbers look convincing. To assure the desired accuracy, we'll compute a simple integral to bound the upper tail:

$$\sum_{k=n+1}^{\infty} a_k \leq \int_{k=n+1}^{\infty} a(x)\, dx.$$

```
In[3]:= Integrate[a[x],{x,n+1,Infinity}]

              1
Out[3]=     -
              n
```

Clearly, $n = 10000$ will do.

Example 2. Find an n for which $\sum_{k=1}^{n} \frac{1}{k^{1.1}}$ differs from $\sum_{k=1}^{\infty} \frac{1}{k^{1.1}}$ by less than .0001.

Solution. First some numbers:

```
In[4]:= a[k_]:= 1/k^1.1 ; Clear[s];
        s[n_]:= Sum[a[k],{k,1,n}]

In[5]:= {s[20],s[100],s[200]}
Out[5]= {3.19146,4.27802,4.69888}
```

Convergence seems to be slow; we'll probably need a large n to assure the desired accuracy. Let's compute the same integral bound:

```
In[6]:= Integrate[x^-1.1,{x,n,Infinity}]

            10.
Out[6]=   ----
            0.1
          n
```

Making this small (i.e., less than .0001) will require a large n as shown below.

```
In[7]:= (10*10000)^10
Out[7]= 10000000000000000000000000000000000
        0000000000000000000000000000000
```

Example 3. Estimate $\sum_{k=1}^{\infty} \frac{1}{10^k}$ to within .0001.

Solution. This series converges much faster than the previous one:

```
In[8]:=Integrate[1/10^x,{x,n,Infinity}]

                1
Out[8]=   -----------------
            n Log[10]
          E           Log[10]
```

If, say, $n = 4$:

```
In[9]:=  % /. n -> 4


                1
Out[9]=  -------------
            10000 Log[10]



In[10]:=  N[%]
Out[10]=  0.0000434294

In[11]:=  Sum[1/10^k,{k,1,4}]
Out[11]=  1111/10000
```

Thus, $S_4 = .11110$ estimates the limit accurately to four decimal places.

Since the given series happens to be geometric, we can compute its limit directly:

```
In[12]:=  1/(1-1/10) - 1
Out[12]=  1/9

In[13]:=  N[%]
Out[13]=  0.111111
```

Thus the estimate is indeed good to the fourth decimal place.

Observe how crucial the integral test was to these calculations.[3] This is as it should be: the close connection between integrals and sums—in particular, improper integrals and infinite sums—is at least as important as any other idea in the area. Using either to approximate the other should help students better understand both.

5.2 Symbolic Computing and Problem Solving

Can symbolic computing be used to improve students' problem solving abilities?

Problem solving can mean many things. Sometimes it means a top-down approach to problems: interpret the problem mathematically, break it into parts, do what needs to be done to each; reassemble the pieces. Symbolic computing might indeed help with some of this, especially with problems that reduce readily to routine symbolic or numerical computations. Graphical computations can also help generate good guesses.

Nevertheless, symbolic computing, in itself, is hardly a panacea for problem solving ills. Symbolic computing solves, directly, only the easy parts of hard problems. For the hardest parts—distilling out the mathematical essence of a problem, recognizing and describing its mathematical structure, and interpreting the results of computation—students, and we, may still be mainly on our own.[4]

In fact, symbolic computing seems to me to be better suited to improving the bottom up style of learning than the top down. That is, it can help students strengthen **basic** mathematical foundations and intuitions: the idea of function, the meaning of convergence, etc. The concrete experience of seeing and manipulating basic mathematical objects in all sorts of guises—graphic, numeric, symbolic, and perhaps dynamic—should help students construct stronger, more flexible, and more useful mental models of these key ideas. Paradoxical as it may seem, high technology will probably do more to refocus undergraduate courses on basic mathematical ideas than to automate, jump-start, or supercharge them.

5.3 Leapfrogging Poor Algebra Skills: a Good Idea?

Can symbolic computing compensate for poor algebra skills? Might it help algebraically ill-prepared students succeed in calculus, and beyond?

A guess (and it is only that) is that while symbolic computing might be useful in helping students **improve** algebra skills, it won't **substitute** for them. Calculus—all of the above notwithstanding—is algebra-intensive. Throughout the subject, algebraic and symbolic intuition is essential. Most of the main objects of calculus (derivative, series, integral, etc.) are defined via algebraic combinations, sometimes quite complicated, of functional expressions. Can the idea of definite integral as limit of Riemann sums, or derivative as limit of difference quotients, possibly become clear, or usable in applications, without a seat of the pants feeling for algebraic notation? I doubt it.

---

[3]We used the integral test only to find upper bounds for upper tails. We could also have found lower bounds.

[4]Symbolic computing is no panacea, but it is hardly irrelevant to the harder parts of problem solving. Consider, for example, Alan Schoenfeld's 4-fold taxonomy of problem solving: Resources, Heuristics, Control, and Belief. It is natural to expect students' experiences with computing to affect all four areas crucially. (cf. Schoenfeld's paper in [6].)

6   Costs and Obstacles

Whatever the benefits of symbolic computing, there are also, undeniably, significant costs. They come in many forms: money, time, aggravation, and diversion.

To make a long and sometimes painful story short: Expect trouble. Machines break, things take longer than expected, and students won't always thank us for our efforts. On the contrary, many calculus students expect little more than routine symbol manipulations in college calculus. A more conceptual approach to the subject often adds up to a harder course. Students accustomed to solving dozens of routine template problems at a sitting are in for a new, and for some, anxiety-provoking experience.

Teaching courses with symbolic computing is not easier, either. In the short run, at least, it's harder, and it takes more time. Many of the utterly routine exercises in standard calculus texts, for example, make no sense for a student equipped with symbolic computing; they call for little more than typing. Creative, open-ended problems, on the other hand, are hard to create: finding the right balance between boring students with cookbook directions and mystifying them with vague generalities is surprisingly difficult.

The moral is: don't despair, but be realistic. Mathematics is hard—to do, to teach, and to learn. The calculus course has, over the years, become easy to teach, but only because we've stripped it down so far, not to its essentials but to its frills. Any effort to improve it, with or without technology, will mean work.

The main point is that the work is worth doing. Symbolic computing raises problems, but they are solvable and, what's more, worth solving.

7   The Need for a Rationale

The tangible needs raised by symbolic computing are obvious: equipment, technical support, course materials, time, and professional rewards.

Less tangible, but just as important, is the need for a **rationale** that is convincing—to us and to our colleagues—for using symbolic computing. When we understand clearly, and state forthrightly, what we hope to gain from symbolic computing, we will find ways through or around the logistical problems computing will inevitably raise. Without a clear rationale, logistical problems will probably prove fatal.

It's important, finally, to keep things in perspective. Some, perhaps even most, educational experiments with symbolic computing will, by some measures, fail.

But compared to what? The best students will always learn, because of or even in spite of our best efforts. The rest will always, by definition, be harder to teach, with or without technology, but how well are they served now?

8   Conclusion

We should use symbolic computing in undergraduate mathematics because, as I've argued, it can help us teach better, and our students learn better, what we really care about: mathematical ideas. With symbolic, graphic, and numeric computing we can show and handle mathematical ideas more concretely, more directly, and from more perspectives.

The title of this paper mentions insight. That word, admittedly, is overused. But here insight may indeed be appropriate: seeing ideas more concretely, more directly, and from more perspectives strikes me as a good working definition.

Nearly ten years ago Herb Wilf [7], writing about *muMath*, sounded a "distant early warning" for freshman calculus. Wilf's warning is apt, but the military image he conjures up—warning sirens, bombers over the horizon—is unnecessarily gloomy. The image of forest fire might do better. Violence is done, but what burns is mostly deadwood and brush. Light and air are admitted, and new growth can occur.

References

1. *Computer Algebra Systems in Education (CASE) Newsletter,* c/o Don Small, Department of Mathematics, U. S. Military Academy.

2. Douglas, R. G. (ed.) (1986). *Toward a Lean and Lively Calculus: Report of the Conference/Workshop to Develop Curriculum and Teaching Materials for Calculus at the College Level.* MAA Notes Number 6, Mathematical Association of America.

3. Smith, D. A., Porter, G. J, Leinbach, L. C., and Wenger, R. H. (eds.) (1988). *Computers and Mathematics: The Use of Computers in Undergraduate Instruction,* MAA Notes, Number 9, Mathematical Association of America.

4. Steen, L. A. (ed.) (1988). *Calculus for a New Century,* MAA Notes, Number 8, Mathematical Association of America.

5. Tucker, T. W. (ed.) (1990). *Priming the Calculus Pump*, MAA Notes Number 17, Mathematical Association of America.

6. *Proceedings of the St. Olaf Conference*, (1989). (12 papers on symbolic computing and undergraduate mathematics), c/o Paul Humke, Department of Mathematics, St. Olaf College.

7. Wilf, H. E. (1982). "The disk with a college education," *The American Mathematical Monthly*, 89, 1, 4–8.

Appendix: A Geometric Look at the Bieberbach Conjecture

Understanding what deep theorems and principles of analysis mean, not just what they say, is always a challenge. Visual representations can be surprisingly effective—even in "hard" analysis areas.

According to the Riemann mapping theorem, every simply connected domain other than the entire plane is conformally equivalent to the unit disk—i.e., equal to the image of the unit disk under a 1–1, analytic mapping. For example, the Koebe function $k(z) = z/(1-z)^2 = z + 2z^2 + 3z^3 + \ldots = z \cdot \frac{d}{dz}(1/(1-z))$ maps the unit disk conformally to the whole complex plane, except for the slit $x \le -1/4$ along the real axis. It's easy to see, too, that $k(0) = 0$, that $k'(0) = 1$, and that k is 1–1 on the unit disk. A *Mathematica* picture illustrates the situation.

```
In[1]:= k[z_] := z/(1-z)^2
In[2]:= PolarMap[k,{0,1},{0,2Pi},
           PlotPoints -> 25]
Out[2]=
```

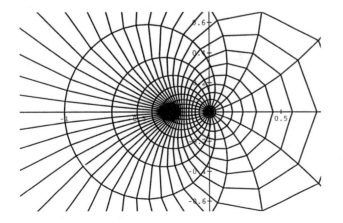

Near $z = 0$, k appears to behave like the identity function—as it should, given that k agrees with the identity to first order at the origin. Near $z = -1/4$, something else seems to be happening. Let's have a closer look:

```
In[3]:= Show[%, PlotRange ->
           {{-.5,0},{-.2,.2}}]
Out[3]=
```

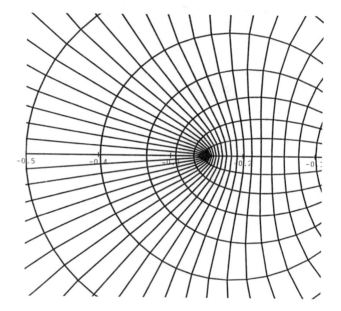

The action at $z = -1/4 = k(-1)$ is no accident; it illustrates Koebe's important "Quarter Theorem." Details can be found in standard complex analysis texts.

The Bieberbach conjecture—now de Branges' theorem—asserts that the function k is, in a strong sense, extremal among all normalized conformal maps of the unit disk:

Theorem 1 (de Branges, 1984) *Suppose $f(z) = z + a_2 z^2 + a_3 z^3 + \cdots$ converges for all complex z with $|z| < 1$, and represents a univalent function in the unit disk. Then for all $n > 1$,*

$$|a_n| \le n.$$

If, for any n, $a_n \ge n$, then f is "essentially" the Koebe function, k.

In other words, the coefficients of a suitably normalized one-to-one analytic function, defined on the unit disk, can't be too large. If even one of the a_n's exceeds n, the function fails to be one-to-one.

The following *Mathematica* pictures illustrate what the theorem says. The proof itself is beyond our scope. First, we'll enlarge slightly the coefficient of z^2 in the power series for k, and plot the result:

```
In[4]:= k1[z_] := k[z] + .1 z^2
In[5]:= PolarMap[k1,{0,1},{0,2Pi},
          PlotPoints -> 25]
Out[5]=
```

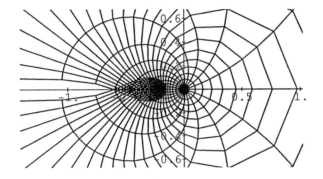

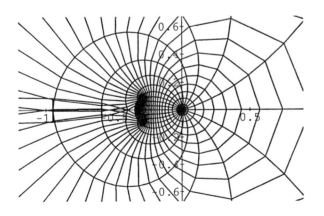

The theorem offers no help here, but the pictures show clearly (and, perhaps, surprisingly) that $k2$ is not one-to-one. Similar perturbation experiments on other coefficients would yield similar results.

As the theorem predicts, $k1$ is not one-to-one. What if we **decrease** the coefficient of z^2?

```
In[6]:= k2[z_] := k[z] - .1 z^2
In[7]:= PolarMap[k2,{0,1},{0,2Pi},
          PlotPoints -> 25]
Out[7]=
```

```
In[8]:= Show[%, PlotRange ->
          {{-.5,0},{-.2,.2}}]
Out[8]=
```

Naturalizing Computer Algebra for Mathematical Reasoning*

Tryg A. Ager
Stanford University

Sam Dooley[†]
University of California, Berkeley

1 Introduction

From our perspective, computer algebra is like an immigrant from computer science to mathematics. It is welcomed as a productive participant, but it may have to overcome a language barrier because native "computer algebra" is not effective or appropriate in all mathematical contexts. So we expect the languages of computer algebra to change in a process of acculturation and naturalization. There are many indications that this process is occurring in a constructive way, but there is always a chance that adaptation may not occur. In South Dakota in the 1970's, some descendants of Norwegian immigrants who still spoke 19th-century dialects were studied by Norwegian linguists who could find no such speakers in Norway itself. We could imagine future computer scientists traveling to distant places to study principles of computer algebra prevailing in the 1990's. We hope such travelers would have the outlook that capabilities for valid, mathematically natural reasoning are essential to computer algebra, and that their interest in any surviving systems lacking such qualities would be purely academic.

The purpose of this paper is to explain our attempt to naturalize computer algebra in an experimental system called *EQD* (because it supports "equational derivations") that is oriented toward valid, interactive mathematical reasoning. *EQD* uses *Reduce* as an embedded algebra engine, and is part of *Dfx* [17, 1, 5], a computer-based system for instruction in elementary calculus. *Dfx* and *EQD* have been used successfully for instruction [16], but in this paper we focus on the general problem of adapting computer algebra to specific mathematical purposes. We begin by identifying components we believe are necessary for a natural reasoning-oriented system.

Characteristics of a Natural System

1. **Its Purpose is to Support Rigorous Interactive Mathematical Reasoning.** Rigorous mathematical reasoning involves explicit step-by-step progression from initial premises to conclusions using certifiably correct transformations. In a computer-based interactive setting, rigorous mathematical reasoning need not interfere with the natural development of a mathematical argument and need not strangle understanding with excessive detail. Interactive reasoning means that the system and the user share the effort over a sequence of steps that lead consistently from assumptions, definitions, or given propositions to derived conclusions. The user decides what to do in accordance with the accepted methods and style of the mathematics being done. The reasoning-oriented system then executes the proposed operation if it finds that the operation produces a valid result, or rejects the proposed operation as incorrect. Strategic direction is determined by the user, but certification and execution of tactics are left to the system.

Other mathematical software systems have a different focus, such as algorithmic programming and interactive evaluation [18, 8] or pattern matching [8, 20] or have adopted metaphors such

---

*Research supported by the National Science Foundation Grants MDR-85-50596, MDR-87-51523 and MDR-90-50008 at Stanford University, and by the National Science Foundation Grants CDA-87-22788 and CCR-88-12843 and the Defense Advanced Research Projects Agency (DoD) under ARPA Order 4871, monitored by the Space and Naval Warfare Systems Command under Contract N00039-84-C-0089 at the University of California at Berkeley.

†Supported under a National Science Foundation Graduate Fellowship.

as module interconnection management [12], interactive notebooks [2, 20], or mathematical expression editing [14, 6, 11]. Compared to these systems, our emphasis is on the everyday concept of mathematical reasoning, where one starts with premises, applies mathematical operations, and by a series of valid steps, arrives at a conclusion.

2. **It is Cognitively Faithful to Conventional Mathematics.** The interface must be natural in the sense that it obeys the notational conventions and inferential norms of the mathematical specialty it serves. This criterion is part of the "no foreign language" requirement, but goes farther in requiring that the system fit well with the way mathematical thinking proceeds.

   Considering the many different specializations and subdisciplines in mathematics, our immigration analogy suggests that as part of its mathematical acculturation, computer algebra must specialize and learn the technical subdialects and reasoning tactics appropriate for specific mathematical subdisciplines. If the system is targeted for instructional use, it is even more important that natural channels of mathematical learning and development be available to the student. Deliberately generic systems like *Reduce* or *Mathematica* are bound to be unnatural for particular problems. However, we will demonstrate that they can be made natural for a specific purpose.

3. **Reasoning is Presented Externally as Derivations.** Recent systems that combine symbolic computation with report preparation facilities such as *FrameMath*'s expression editing [6], *MathCad*'s blackboard [10], *CaminoReal*'s interactive documents [2], and *Mathematica*'s notebooks [20] all go a long way toward satisfying this requirement. Appearance is necessary for the naturalization of computer algebra, but it is not the whole story. From a reasoning-oriented perspective, appearance is the least important of the defining properties.

4. **Reasoning is Represented Internally as Logically Valid Derivations.** There must be internal data structures that represent derivations and internal procedures that validate them. It is one thing for formulas to look as if they follow from one another, but quite another thing to certify that they do. An essential property of a reasoning-oriented system is that it completely implements the idea of inference as a series of logically valid steps, checking that formulas follow from one another according to certifiable transformations. Most computer algebra systems strive to be correct, but because they provide little or no representation of inference, there are many circumstances where their results are mathematically wrong.

No system, including *EQD*, yet achieves total correctness. Because *EQD* was developed for the limited domain of univariate calculus and its intended users were students instead of mathematical experts, we focused on correctly handling the troublesome aspects of elementary calculus (singularities and other pathological behavior of functions). To be instructionally effective, we needed a correct system because we cannot depend on students' judgements to filter incorrect results. For example, when *Mathematica* is asked to differentiate $(-2)^x$, it will return something equivalent to $\text{Log}[-2]*(-2)^x$, since it performs computations over the complex numbers. *EQD*, which checks for differentiability over $\Re$, is able to inform a student that the function is not differentiable. Pursuing correctness for pathological cases led to a system of representing the semantics of elementary functions, which is discussed more fully in [1, 5], and it led to innovative utilization of pattern-driven algebra, which is the primary technical focus of this paper.

These requirements are formidable, and one question that arises is whether or not to start from scratch to satisfy them. Mathematically, the two main issues are that elementary univariate calculus is constrained to the reals, and the validity of the limit, differentiation, and integration operations depends on properties of intervals on which functions are defined. On both points, *Reduce* acted incorrectly or inappropriately and needed to be naturalized. But even though it is a generic system, *Reduce* is modular enough that it could be embedded in a reasoning-oriented interface. The advantages of *Reduce*'s algebraic algorithms (Risch integration and Hensel factorization, for example) were thereby retained. Treatment of its deficiencies for calculus instruction are described in subsequent sections.

2   Motivating Examples

We now present several motivating examples that shaped our outlook on construction of a reasoning-oriented interface.[1] The substantive mathematical issues about intervals on $\Re$ are actually buried in the validity maintenance and inference checking routines and are not controversial as to their specification, although they are complex from a programming standpoint. But naturalness for the special area of calculus instruction is also strongly related to less precisely specifiable aspects, which we group together as notational conventions and inferential norms.

Example 1. We begin with a true story. We once did a prototype for teaching calculus, DFDEMO, that placed a thin reasoning shell around *Reduce*. A mathematician who saw DFDEMO was seriously annoyed because the input `f(x) = x + a` came back as `f(x) = a + x`. He expressed equal annoyance that `g(x) = sqrt(x**2)` was returned as `g(x) = x`. He felt the first was annoying because reordering was gratuitous and unnatural. The latter was just wrong. Despite all we did to try to explain the former as a trade-off required by *Reduce* for automatic simplification, and the latter as a pitfall of pattern-driven algebra, he remained convinced that mal-ordering and mal-computing were equally pernicious. His objections arose because he perceived DFDEMO as a medium for presenting sacred material to students in a natural way, instead of as a private tool for doing computations.

Example 2. Now consider the algebraic operation of clearing roots in a fraction such as $(\sqrt{x+h} - \sqrt{x})/h$. This kind of problem turns up in taking the limit of the difference quotient of a function in order to compute its derivative. Figure 1 shows how *Reduce* either does nothing or trivially solves the problem, depending on how it is presented.

Someone planning to rationalize the numerator may reason that multiplying through by the conjugate will clear roots and cancel h. However, *Reduce* left to itself derails this line of thought if an unfortunate stenographic choice is made. Such brittle behavior, where a slight notational change in the input produces a large inferential difference in the output, will immensely frustrate users of symbolic computation systems, just as numerical routines with similarly anomalous input/output behavior undermine users of scientific programming applications.

In contrast, Figure 2 demonstrates how a natural reasoning-oriented interface supports the proposed so-

---

[1]All examples in this paper involving *Reduce* use Version 3.3, released on 15 January 1988.

```
VM/Reduce 3.3
% Reduce goes nowhere if the conjugate is formed
% as a separate quotient of the form (a/h) * (c/c)
1: ((sqrt(x+h) - sqrt x)/h) * (sqrt(x+h) + sqrt x)
   /(sqrt(x+h) + sqrt x);

 SQRT(H + X) - SQRT(X)
-----------------------
           H

% But given a quotient of products of the form
% (a * c)/(h * c), Reduce clears roots in the
% numerator and cancels all at once.
2: ((sqrt(x+h) - sqrt x)*(sqrt(x+h) + sqrt x))
   /(h*(sqrt(x+h) + sqrt x));

            1
-----------------------
 SQRT(H + X) + SQRT(X)
```

Figure 1: Reduce output for clearing roots.

lution attempt, even though all algebraic operations are still being done by *Reduce*. When we claim that the derivation in Figure 2 is produced within a natural, reasoning-oriented system, we mean specifically that:

- It is developed interactively. This derivation has one premise that defines a function. Then the closed-form representation of the function is directly manipulated to derive a form with roots cleared in the numerator.

- It is cognitively faithful to mathematics. Literally, we defined a function, multiplied numerator and denominator by a new term, factored, collected terms and canceled. Note also that **equations**, not just terms, are present throughout, adding to the mathematical faithfulness.

- It looks like a derivation. Some would dispute this because of the typescript presentation. But a screen presentation would show just step 1 and the current state of step 2, which changes from one algebraically equivalent form to another much like the equality chain shown below.

- It has the internal structure of a valid derivation. Note that internally *EQD* has figured out that the terms in the square roots must be positive and that the symbolic term h cannot be 0. Step 2 adds more restrictions, which are required because division by a new term occurred. *EQD* uses these restrictions to maintain validity.

To reinforce the point that the derivation in Figure 2 is natural, consider the following three other ways to express the reasoning involved.

```
Dfx/EQD, 09/26/90 17:52:10
% Begin by defining our function
calc> define f(x) = (sqrt(x+h) - sqrt x)/h
            +-----+   +-+
            \|x + h  - \|x
   1. f(x) = ----------------
                    h
       provided that h + x >= 0, x >= 0, h <> 0.
% Multiply numerator and denominator (mnd)
% by the conjugate of the numerator
calc> 1 mnd @ 2; sqrt(x+h) + sqrt x
            +-----+   +-+   +-----+   +-+
           (\|x + h  - \|x )*(\|x + h  + \|x )
   2. f(x) = ---------------------------------
                      +-----+   +-+
                    h*(\|x + h  + \|x )
       provided that h + x >= 0, x >= 0, h <> 0,
       +-----+   +-+
       \|h + x  + \|x  <> 0.
% We multiply out (mo) ...
calc> 2 mo @ 2 1
                      2
        +---+   +---+  +-+  +-+  +---+  +-+  +-+
       \|x+h  +\|x+h *\|x -\|x *\|x+h -\|x *\|x
   2. f(x) = ---------------------------------------
                      +-----+   +-+
                    h*(\|x + h  + \|x )
% ... which updates step 2 in place.
% After collecting exponents and canceling
% terms in the numerator, [omitting these
% details for brevity] we arrive at
                        h
   2. f(x) =  -------------------
                 +-----+   +-+
               h*(\|x + h  + \|x )
% And choose to cancel the factor of h
calc> 2 cancel @ 2; h
                      1
   2. f(x) =  ---------------
               +-----+   +-+
               \|x + h  + \|x
```

Figure 2: *EQD* output for clearing roots.

First, a discursive sketch (description) of the reasoning is that multiplying numerator and denominator by a conjugate will clear roots by elevating powers in a way that allows extraction of h as a factor of the numerator, which then cancels with the factor of h in the denominator.

Second, a procedural sketch (recipe) of the same reasoning is: first, multiply numerator and denominator by the conjugate of the numerator, second, simplify the numerator, and third, cancel h from numerator and denominator.

Finally, a textbook presentation might use this equality chain:

$$\frac{\sqrt{x+h}-\sqrt{x}}{h} = \frac{(\sqrt{x+h}-\sqrt{x})(\sqrt{x+h}+\sqrt{x})}{h(\sqrt{x+h}+\sqrt{x})}$$
$$= \frac{(x+h)-x}{h(\sqrt{x+h}+\sqrt{x})} = \frac{1}{\sqrt{x+h}+\sqrt{x}}$$

where each step is taken from a set of well-defined

operations that could have been described earlier in the text.

Note that all these descriptions are natural. The idea of naturalness includes redundancy and alternative derivations of the same result. The rigor of derivation and proof does not imply a straitjacket constraining how derivations proceed. In *EQD* it is possible to execute the same general strategy in other ways than the ones shown above: a user can have the system confirm the conjecture that h is a factor of the numerator, or can have the *Reduce* factorizer extract factors, which would include h, or have *Reduce* confirm that the numerator simplifies to h.

For any given segment of mathematics, there is hardly ever just one way of doing things, and a natural interface must cover the generally accepted derivations, not just one distinguished method. This requirement complicates the effort to systematically identify interface components and implement them coherently, a subject we return to in section 3.

Example 3. Finally, we present an example of mathematical reasoning found in the theoretical development of elementary calculus. In this case we are using the system to justify a general rule for differentiating positive integral powers of x with respect to x. The derivation assumes the basis for induction, $dx/dx = 1$, and the availability of the product rule. Shown in Figure 3, the proof is very much like what is found in elementary textbooks, and is presented in Review format, where the user interaction is omitted.

Often in elementary calculus instruction, there is not a strong distinction between methods of derivation used for solving problems and those used to extend the theory. Like the integral power law "proved" in Figure 3, many elementary proofs are just the implicitly general algebraic core of fully articulated proofs found in treatises on analysis. *EQD* reflects this easy transition between problems and proofs, both in the granularity of the available operations and with respect to the established inferential norms of elementary calculus.

Highlights of the derivation in Figure 3 are as follows: Step 1 defines a function: the user types **define f(x) = x**n. *Reduce* verified it was a function on $\Re$. Steps 2 and 3 introduce premises for the induction; *Reduce* verified both were meaningful on $\Re$. Step 4 came by locating a target in step 1 and typing in the replacement term **x*x**(n-1)**. *Reduce* certified the replacement proposal. Applying the product rule to step 4 by typing **4 dproduct x** produced step 5. Re-

```
Dfx/EQD, 09/26/90 17:52:10
                n
1.  f(x) = x

We have the base case for an induction
      dx
2.  -- = 1
      dx

And the induction hypothesis
      d  (n - 1)                (n - 2)
3.  --(x         ) = (n - 1)*x
      dx

Step 1 is algebraically equivalent to
               (n - 1)
4.  f(x) = x*x

Then differentiating using the product rule
            d  (n - 1)       (n - 1) dx
5.  f'(x) = x*--(x       ) + x       *--
            dx                        dx

>From the base case we replace dx/dx
            d  (n - 1)       (n - 1)
6.  f'(x) = x*--(x       ) + x       *1
            dx

>From the induction hypothesis (step 3)
we replace the other derivative
                     (n - 2)    (n - 1)
7.  f'(x) = x*(n - 1)*x       + x       *1

And simplify the result (details omitted)
               (n - 1)
8.  f'(x) = n*x
```

Figure 3: *EQD* derivation of integer power law.

placing equals by equals using steps 2, 3, and 5 and some interactive simplification obtains the final result. Remarks were inserted along the way and are parts of the internal representation of the derivation.

All the algebra and checking in the derivation in Figure 3 is done by *Reduce*, but the output does not look or act like *Reduce*, nor did the user use a *Reduce* program to produce derivations. Instead the user made natural conjectures about how to solve the problem in the form of easily articulated steps that may or may not lead to the solution. The reasoning partnership is illustrated here in that: (1) Acceptance of this derivation as a proof of the integral power law depends on interpreting the premises (steps 2 and 3) as the correct hypotheses for the induction theorem. The induction theorem itself is not implemented in *EQD*. (2) Strategy selection is a user's responsibility. (3) Reduce is responsible for correctly executing a strategy component under the control of *EQD*, which certifies the correctness of the derivation by certifying that each derived step preserves validity.

Various things can hinder a proof attempt, such as mistyping a formula, trying an inappropriate operation or plan, or being at a loss about what to do next. A positive quality of a reasoning-oriented interface is that there is not much of a gap between what

you know about the subject and what you can do in the system. Conversely, difficulties stem from deficient mathematical understanding, not from inability to translate natural solutions into computer idioms.

3 Naturalization Techniques

In this section, we will discuss the techniques that allowed us to present a natural reasoning-oriented interface around *Reduce*. The fundamental programming issue was designing a data structure that represented the details of valid inferences, allowed the natural construction of derivations, and retained the ability to communicate with *Reduce* and its general programming language. Here we focus on the main techniques we used for naturalizing *Reduce*'s pattern matcher and its built-in simplification functions, and present principles we used to organize algebraic operations from a reasoning-oriented perspective.

3.1 Difficulties with Pattern-Matching and Simplification

Many natural operations in elementary algebra are best expressed as declarative theorems or formulas that lend themselves to pattern-based transformations. Other properties which are not troublesome for students, such as associativity, commutativity, and distributivity, cause nothing but problems for the automated pattern-based approach. Students routinely learn to use declarative formulas, and several computer algebra systems allow the user to define pattern-based transformations. Like other computer algebra systems, *Reduce* has a means (**FOR ALL...SUCH THAT...LET...**) for specifying the substitution of one expression by another. The substitution may involve pattern variables and conditional application. (For a more complete description see [8].) The *Reduce* 3.3 pattern matcher is neither the most modern nor the most powerful available, but as with fine wines, carefully selected older vintages with the right accompaniments make excellent fare. In order to serve up a more natural interface we restructured the *Reduce* pattern matcher in several areas to accommodate the natural granularity of pattern-based reasoning in elementary calculus.

Locality. *Reduce* normally applies each rule that the user declares (using the **LET** command) to all of the expressions it simplifies until the rule is explicitly removed (using the **CLEAR** command) by the user. Unfortunately, requiring a student to explicitly control which patterns are used for simplification is undesirable since the concept of a set of "globally declared patterns" arises from the foreign language of the computer algebra system. However, the set of globally active patterns must be changed frequently to prevent the computer algebra system from performing too much or too little simplification, as was illustrated in Figure 1. To obtain more natural control over which patterns are applied to an expression, we implemented an internal mechanism for localizing pattern application. It automatically installs a group of rules, evaluates an expression, and finally restores the set of patterns to a well-defined state.[2] The multiplication of numerator and denominator by $\sqrt{x+h}+\sqrt{x}$ shown in Figure 2 used this technique.

Repetition. *Reduce* repeatedly applies patterns until the input expression no longer matches any installed pattern. If a pattern matches its own consequent, this repetition causes problems because the matching process will not terminate. For example, applying the commutative law to $x + y$ yields $y + x$, which yields $x + y$, and so on. In addition, any useful mathematical fact that can be stated as an equation can be translated into two transformations that operate in opposite directions. For example, the distributive law yields the patterns $a(b + c) \Rightarrow ab + ac$ and $ab+ac \Rightarrow a(b+c)$. Yet if both directions of the identity are naively translated into pattern-based rewrite rules and installed simultaneously, the pattern-matcher will again fail to terminate. If only one is installed by default, problems that require the other become unnatural. In order to avoid forcing users to resort to the control language of computer algebra, we built a mechanism that allows a pattern to be applied exactly once.[3] This gives us a controlled way of directly using the standard identities found in any algebra or trigonometry text, avoiding the circularities described above.

Ambiguity. Often an expression can be transformed by more than one pattern. For example, when collecting terms in the expression $x^2 x^2$ using exponent operations, should the result be $(x*x)^2$ or x^{2+2}? Both the rule $a^c b^c \Rightarrow (ab)^c$ (which collects common exponents over different bases) and the rule $a^b a^c \Rightarrow a^{b+c}$

(which collects common bases having different exponents) can naturally be described as a law of exponents that collects terms, and each can be applied to the given expression. Any well-defined automatic collection operation using these rules must resolve this ambiguity in favor of one rule or the other. No matter which one is chosen, someone will soon need the other one. In such a situation it is often most natural to allow the user to make the choice by providing the capability for one-time application of each rule independently.

Selective Application of Simplification Functions. *Reduce*'s preprogrammed simplification functions (the `SIMPFNS`), like those supplied in other computer algebra systems, are algorithmic, optimized simplifiers for common operations such as addition, multiplication, integration, etc. Simplification functions can be a major cause of unnatural behavior with respect to the style of algebraic reasoning used in elementary calculus, because algorithms that can be efficiently programmed often lack a natural conceptual model. This problem motivates the interactive use of less powerful rules in the first place. Also, since the simplification functions and the pattern-matcher often alternate during evaluation, the simplifier has an annoying tendency to interfere with the intended result of applying a pattern. To shield from this interference, in *EQD* we either disable simplification functions or mask the data when using one-time pattern application. For example, by internally translating (PLUS A (TIMES B C)) to (H-PLUS A (TIMES B C)) before calling the evaluator, *EQD* prevents *Reduce*'s addition `SIMPFN` from running at the top-level of the expression. In this way the sum is preserved while the product is fully simplified. Selectively disabling `SIMPFNS` by internally masking data is particularly useful, for example, in allowing students to proceed through the step-by-step differentiation of algebraic expressions using *EQD* commands for the standard differentiation laws. If *Reduce* were allowed to simplify the derivative, very often it would completely solve the problem for the student. In that case, a student could neither generate and execute a solution plan nor discover that the derivative laws work by decomposing complex problems into simpler ones.

Interference with Simplification Functions. The formula $\frac{d}{dx}(\tan x) = \sec^2 x$ is common in textbooks, but *Reduce* uses the equivalent formula $\frac{d}{dx}(\tan x) = 1 + \tan^2 x$ to express the derivative of the tangent. Any sophomoric programmer, using pattern

---

[2] Similar facilities to localize patterns have since been made available through the Reduce Network Library and are included in *Reduce* 3.4.

[3] This mechanism is functionally similar to *Mathematica*'s `ReplaceAll` (`/.`).

rules, can apparently make Reduce conform to the textbooks. However, *Reduce*'s implementation of the Risch integration algorithm requires that the derivative be expressed in terms of the tangent, so expressions appearing in further steps of the algorithm can be assumed to be in a certain standard form. If the derivative of the tangent is modified, the integrator no longer functions correctly. Although it is easy to avoid this particular problem, it points out a major drawback of relying heavily on pattern-based programming: a seemingly harmless pattern can have unexpected and sometimes devastating effects on other components of the system, whether they be procedurally-based functions or other declaratively-based, pattern-driven simplification routines.

Conditional Transformations. When we speak of a conditional transformation, we mean one that applies only when certain restrictions are met by the terms being modified. For example, consider the following law of exponents: $\forall x, a, b \in \Re, x > 0 \Rightarrow (x^a)^b = x^{ab}$. During the course of algebraic simplification, the restriction $x > 0$ should be established before using the above equation to insure that a valid inference will result. However, if the value of x is not known when the transformation is to be applied, it may not be possible to do so. Traditional computer algebra systems have taken two approaches to this problem. On the one hand, a system can ignore the restrictions and apply the pattern anyway. Several of *Mathematica*'s built-in simplification functions exhibit this behavior, one of which (`PowerExpand`) asserts that $(x^a)^b = x^{ab}$ in all cases. Allowing such invalid transformations is inadequate for a system that attempts to certify mathematical reasoning. On the other hand, a system can refuse to apply a conditional transformation at all when its restrictions cannot be established. The conditional patterns provided by both *Reduce* and *Mathematica* (among others) perform in this fashion. This approach is more conservative, but it hobbles the system's ability to use the mathematical knowledge embedded in conditional patterns.

The reasoning-oriented nature of *EQD* allows us to offer a third approach. When a user requests a conditional transformation, *EQD* applies the transformation, carefully notes the restrictions that certify that the transformation is correct, and maintains the restrictions alongside the result of the transformation. In this way the standard problem of what to do with $\sqrt{x^2}$ can be correctly handled in a natural way. If the user asks that this term be simplified to x, the restriction that $x \geq 0$ will be maintained with the result during further steps in the derivation. Alternatively, the user can ask that $\sqrt{x^2}$ be simplified to $|x|$, which incurs no such restriction.

Simplification Control. Pattern-matching provides flexible control over low-level algebraic manipulations. However, in systems with little more than this capability, such as *FrameMath*, solving more complicated problems requires a large number of user interaction steps. In contrast, powerful computer algebra systems that attempt to automate all of mathematics tend to remove control from the user by not providing convenient enough access to elementary algebraic principles. (See the examples from section 2.) Convenient user control of low-level simplification is particularly lacking in *Mathematica*, where internal patterns that cannot be changed often interfere with the natural development of mathematical reasoning.[4]

From the standpoint of natural mathematical reasoning, some steps are computationally intensive but easy to conceptualize, while many others are merely straightforward transformations on expressions. In order to deliver accurate control over algebraic simplification, a reasoning-oriented system must provide facilities (whether derivational rules, pattern-based transformations, or command-driven simplifications) that span a wide range of computational power, from elementary properties of algebra to computationally intensive algorithms like the Risch integration algorithm. Providing the appropriate level of control over algebraic simplification motivated the design principles presented in the following section.

3.2 Principles for Development of a Natural System

Providing the right degree of control in an interactive computer-based system is analogous to finding the right level of conceptual detail in off-line mathematics. We do not expect that modifications to the pattern-matcher and simplification functions will fully naturalize any given computer algebra system. A mathematically natural system depends both on supplying controllable computational power and principled analysis of the mathematical subject matter. In

---

[4]Examples vary from version to version: in *Mathematica* 1.2, the pattern `Sec[x] := 1/Cos[x]` is always applied; in version 2.0, the reverse of the above pattern is always applied. Either way, control is gratuitously taken away from the user.

our work on *EQD*, we went through many phases in attempting to organize the natural mathematical operations needed for elementary calculus. We finally abstracted these fundamental principles for designing a natural reasoning-oriented system:

1. **Foundations.** Implement a standard theoretical foundation completely. A reasonable foundation for a system intended for calculus instruction and problem-solving begins with the algebra of rational functions over the reals.

2. **Extensions.** Specify and implement new operations as consistent extensions of the foundations, but organize and classify extensions in cognitively efficient ways. We want to insure that the cognitive load on students increases minimally as the mathematical content is extended.

3. **Consolidation.** Combine and generalize sets of mathematically simpler operations. Add algorithmic operations as task-oriented "power tools" that operate within the framework of the theoretical foundation. Consolidated operations reformulate foundational operations, without increasing the theoretical power of the system, so the capabilities of the system can be more effectively used in complex problem-solving situations.

4. **Redundancy.** Implement redundant natural interfaces to mathematically equivalent operations. We have observed that user preferences often vary among mathematically equivalent techniques; the redundancy principle grants the legitimacy of these variations.

We found that using these principles to organize the basic operations of an algebraic system helped us provide a more complete and uniform set of capabilities and helped us create a natural reasoning-oriented interface. The following paragraphs discuss these principles in greater detail.

Foundations. Each operation in a reasoning-oriented system should be based on a consistent, formal mathematical theory. Conversely, the foundations should be explicitly available as operations in the system. Standard formal theories make good starting points for implementation specifications because they frequently facilitate natural reasoning in a given area. For elementary algebra, the foundations include the standard properties of identities, inverses, associativity, commutativity, and distributivity. Each

of these properties is directly implemented by interactive operations in *EQD*. Associativity is implemented by operations for introducing and removing parentheses. Commutativity is implemented by operations for exchanging terms within addition, multiplication and equality. A distributive law such as $a(b+c) = ab+ac$ is implemented by an operation for expanding expressions, $a(b+c) \Rightarrow ab+ac$, and an operation for collecting expressions, $ab + ac \Rightarrow a(b + c)$. The student has the option of choosing any of those operations whenever desired. As shown in Figure 2, foundational operations are often necessary to control the direction of problem solving.

Extensions. Extensions to interactive systems pose a complicated organizational problem, because new operations become intertwined with old ones. There are literally hundreds of algebraic and calculus operations available for problem-solving by the end of the first calculus course. However, analogies between different functions can be used to guide the construction of operations on expressions using analogous operators, minimizing the cognitive load on the user. For example, the distributive operations can easily be extended to include exponentiation: $a^{b+c} = a^b a^c$ is just as much a distribution rule as $a(b + c) = ab + ac$. Alternatively, this identity can be clustered with other exponent rules. Having rule groups exclusively for exponentiation has pedagogical benefits for elementary calculus, where transcendental functions are studied topically. Furthermore, it also brings out similarities between the algebra of logarithms and that of exponentiation while maintaining a degree of modularity between the two sets of rules.

Functionals (limits, derivatives, integrals) are naturally organized by the principal operation of their arguments. Thus we have constant, linearity, product, quotient, and composition (chain) laws for limits, derivatives, and integrals. The strong analogies between operations provided for simplifying functional expressions make it easy to teach the students to use the operations in an interactive reasoning system. The same analogies enable predictable extensions as new functionals are added. A detailed example of how these analogies can be used to guide the construction of operations for manipulating summations (both definite and indefinite) is shown in Section 4.

Consolidation. While we recommend building on theoretically adequate and interactively accessible foundations, we oppose forced labor at pedantically primitive levels. Theoretically accessible resources must be consolidated into single operations in order to exploit computational power. Programs for factoring, solving equations, differentiation, and integration represent typical consolidated operations provided by computer algebra systems. However, these consolidated operations also come at a price: the results may not exactly match what was intended, since control over the computation must be relinquished to the computer. The ability to use foundational operations in concert with consolidated operations allows the student to produce natural derivations.

Because consolidated rules are so powerful, there are inferential contexts in which they are completely inappropriate. Look back at Example 3, the proof of the integer power rule of differentiation. If the consolidated *EQD* rule, DMAGIC, which runs the Reduce differentiator, were used on step 4, it would return the oversimplified term (x**n * n)/x. Worse yet, DMAGIC would beg the question because DMAGIC subsumes the integer power law, yet was used to "derive" the integer power law. If we want to use computer algebra as an aid to proving anything, it must be absolutely certifiable that the internals are not question-begging corruptions of apparently slick external "derivations." Systems designed according to the principles that require both primitive and consolidated methods comply with this demand by allowing us to employ conventional means when consolidated rules would be inappropriate.

Redundancy. We have pointed out that there is seldom only one way to derive a result. Consolidated operations give us one way of providing redundant operations, as do multiple names for the same operation ("multiply out" versus "expand"), as well as completely different alternative methods ("integration by parts" versus "change of variable"). Another type of redundancy is provided by *EQD*'s conjectural rules. Following an idea developed for logic proof checkers [15], a conjectural rule in *EQD* asks Reduce to confirm a relationship between a user-provided term and a target term. So, for example, the ALGEBRA rule checks that the input term is algebraically equal to the target term, and if so it replaces the target by the input. The FACTOR rule uses algebraic division to verify that the input term is a factor of the target term, and if so it replaces the target by the result of factoring out the input. Unfortunately, the conjectural rules in *EQD* are not necessarily complete since they inherit the limitations of the algorithmic simplifiers upon which they depend. Although they may fail to

allow certain correct operations, they are very useful rules in practice.

Summarizing the Principles. These principles are designed to assist students who are learning mathematics. Foundational operations reveal the detailed workings of the mathematics, and allow students to proceed in a tentative and exploratory way until they can formulate a plan. Consolidated operations assist students who have a clear goal in mind and are willing to allow the computer to manage the details. These principles allow us to accomplish the "black box/white box" symbiosis advocated by Buchberger in [3].

Even though these points are being made in the context of algebraic calculations in elementary calculus, they generalize to higher-level mathematics and more sophisticated mathematicians.

4 Extensions to the Reasoning Partnership

Now we want to return to how *EQD* can be a partner in a situation that arises when students encounter a family of similar problems and try to derive a generalization to cover them. Beginning with our earlier example concerning rationalization of radicals in the numerator, assume that a bright student has also worked through the definition of the derivative for $\sqrt[3]{x}$ and $\sqrt[4]{x}$, and begins to see a pattern in the formulas being used to rationalize the numerator. If the pattern could be formalized and proved, it could be used to derive the derivative of the function $f(x) = \sqrt[n]{x}$. The pattern relies on the fact that $\forall n, (a - b) \mid (a^n - b^n)$; the quotient provides the form for the conjugate to use when rationalizing $\sqrt[n]{x - h} - \sqrt[n]{x}$ by taking $a = \sqrt[n]{x - h}$ and $b = \sqrt[n]{x}$. Such a bright student might just see the general formula for the correct conjugate and want to verify it. The conjugate to use, of course, for

$$f(x) = \sqrt[n+1]{x} \text{ is } \sum_{i=0}^{n} a^i b^{n-i};$$

to make sure that the numerator will be correctly rationalized by multiplying by this conjugate we must verify that

$$(a - b) \sum_{i=0}^{n} a^i b^{n-i} = a^{n+1} - b^{n+1}.$$

The proof proceeds as shown in Figure 4.

$$(a - b) \sum_{i=0}^{n} a^i b^{n-i}$$

$$= a \sum_{i=0}^{n} a^i b^{n-i} - b \sum_{i=0}^{n} a^i b^{n-i} \tag{1}$$

$$= \sum_{i=0}^{n} a a^i b^{n-i} - \sum_{i=0}^{n} b a^i b^{n-i} \tag{2}$$

$$= \sum_{i=0}^{n} a^{i+1} b^{n-i} - \sum_{i=0}^{n} a^i b^{n+1-i} \tag{3}$$

$$= \sum_{j=1}^{n+1} a^j b^{n+1-j} - \sum_{i=0}^{n} a^i b^{n+1-i} \tag{4}$$

$$= \sum_{i=1}^{n+1} a^i b^{n+1-i} - \sum_{i=0}^{n} a^i b^{n+1-i} \tag{5}$$

$$= \left(\sum_{i=n+1}^{n+1} a^i b^{n+1-i} + \sum_{i=1}^{n} a^i b^{n+1-i} \right)$$
$$- \left(\sum_{i=1}^{n} a^i b^{n+1-i} + \sum_{i=0}^{0} a^i b^{n+1-i} \right) \tag{6}$$

$$= \sum_{i=n+1}^{n+1} a^i b^{n+1-i} - \sum_{i=0}^{0} a^i b^{n+1-i} \tag{7}$$

$$= \left[a^i b^{n+1-i} \right]_{i=n+1} - \left[a^i b^{n+1-i} \right]_{i=0} \tag{8}$$

$$= a^{n+1} b^{n+1-(n+1)} - a^0 b^{n+1-0} \tag{9}$$

$$= a^{n+1} b^0 - a^0 b^{n+1} \tag{10}$$

$$= a^{n+1} - b^{n+1} \tag{11}$$

Figure 4 : A summation derivation in a
natural extension of *EQD*.

This result cannot be obtained as a single, isolated calculation, and does not make much sense as a program. The proof depends on algebraic detail that is below the threshold of some computer algebra systems, but is entirely natural and appropriate from the point of view of elementary calculus. Highlights of this proof are as follows: (The italicized comments indicate summation operations which are not yet implemented, but which are analogous to already implemented commands for integration.)

1. Elementary distribution operation.

2. *Linear property of the summation operation. A similar operation extracts constant coefficients of integrands.*

3. Fine tuning of exponents. The term a can be changed to a^1 by the identity property, then exponent collection combines the terms. The term b is manipulated similarly.

4. *Change of variable in the first summation. This operation is again completely analogous to the change of variable operation for integration, which requires specification of the change of variable relation (here $j = i + 1$) and eliminates the previous integration variable (here i) from the integrand.*

5. *Renaming a bound variable in a summation. This substitution is a special case of change of variable. Here j is merely renamed to i.*

6. *Splitting the range of summation. This operation is analogous to splitting the range of a definite integral.*

7. Canceling terms.

8. *This is a special summation evaluation formula applicable when the lower and upper limits are the same. There is a corresponding operation for definite integrals.*

9. Computation of the evaluation bar.

10. Canceling terms.

11. Simplification using identity operations.

Although this proof requires operations on summations that are not currently present in *EQD*, it suggests how *EQD*, or any other reasoning-oriented interface, could be extended by adding primitives for the summation operation based on its customary systematization and conceptualization.[5] There is a systematic analogy between summations and definite integrals, which gives coherence and naturalness to a system of operations for one that is extended to include the other.

While under normal circumstances the above problem could not be posed to the average student, it illustrates how *EQD* can assist the exploratory nature of mathematical instruction: as students mature into

---

[5]Operations for manipulating summations have also been proposed by Smith, among others, for "the simulation of hand-manipulation of mathematical expressions in an interactive computer environment" [13]. However, those operations did not have the support of a reasoning-oriented interface, where the proof can be presented as a coherent derivation.

mathematicians, they should have reasoning environments that will accommodate their natural progression from performing rote computation to creating more insightful derivations.

5 Conclusions

This paper describes how we implemented a reasoning-oriented interface called *EQD* that uses *Reduce* to support mathematical reasoning in *Dfx*, an intelligent tutoring system for elementary calculus. The examples in this paper illustrate deficiencies in current computer algebra systems and motivate the approaches we took in designing facilities that correspond to the natural development of the mathematical material. These results bear on several recognized problems for computer algebra, including: how to organize the capabilities of a system, how to grant users more control of both pattern-driven and procedural operations, and how to elevate a user's confidence in a given result by ensuring that a sequence of operations forms a valid derivation. All of these measures are needed to bring computer algebra into the arena of mathematical reasoning as a productive, natural participant. [6]

References

1. Ager, T. A., Ravaglia, R., and Dooley, S. (1989). "Representation of inference in computer algebra systems with applications to intelligent tutoring," In Kaltofen, E. and Watt, S. M., (eds.), *Computers and Mathematics*, Springer-Verlag, New York, 215–227.

2. Arnon, D. S., Beach, R., McIsaac, K., and Waldspurger, C. (1988). CaminoReal: "An interactive mathematical notebook," In [19], 1–18.

3. Buchberger, B. (1990). "Should students learn integration rules?" *SIGSAM Bulletin*, 24, v. 1, 10–17.

4. Char, B. W., (ed.) (1986). *Proceedings of the 1986 Symposium on Symbolic and Algebraic Computation*, New York, 21–23 July 1986. ACM

[6] We would like to thank Patrick Suppes, Raymond Ravaglia, Robert Maas, and Dennis Arnon for stimulating discussions on computer algebra and its implementation; and Richard Fateman for comments on an earlier draft.

SIGSAM, Association for Computing Machinery. Waterloo, Ontario.

5. Dooley, S. (1988). *The use of domain restrictions in computer algebra systems*, Master's thesis, Computer Science Division, EECS Department, University of California, Berkeley.

6. Frame Technology Corporation. (1990). *Using FrameMath*. Frame Technology Corporation, San Jose, California.

7. Gonnet, G. H., (ed.) (1989). *Proceedings of the 1989 International Symposium on Symbolic and Algebraic Computation*, Baltimore, Maryland, 17–19 July 1989. ACM SIGSAM, ACM Press. Portland, Oregon.

8. Hearn, A. C. (1987). REDUCE user's manual, Version 3.3. Report CP 78, The RAND Corporation.

9. Lecarme, O. and Lewis, R., (eds.) (1975). *Computers in Education: Proceedings of the IFIP Second World Conference*, Amsterdam, 1975. North-Holland. Marseille.

10. MathSoft, Inc. (1987). *MathCAD 2.0*. MathSoft, Inc., Cambridge, Massachusetts.

11. Prescience Corporation. (1990). *Theorist: Reference Manual*. Prescience Corporation.

12. Purtilo, J. M. (1989). Minion: "An environment to organize mathematical problem solving," In [7], 147–154.

13. Smith, C. J. (1984). "Implementation of a package of tools for manipulation of sums," *SIGSAM Bulletin*, 18, v. 1, 20–24.

14. Smith, C. J. and Soiffer, N. M. (1986). MathScribe: "A user interface for computer algebra systems," In [4], 7–12.

15. Smith, R. L., Graves, W. H., Blaine, L. H., and Marinove, V. G. (1975). "Computer-assisted axiomatic mathematics: Informal rigor," In [9].

16. Suppes, P. and Ager, T. A. (1991). *Computer-based calculus for high schools: Final report*, Technical Report 316, Institute for Mathematical Studies in the Social Sciences, Stanford University.

17. Suppes, P., Ager, T. A., Berg, P., Chuaqui, R.,
    Graham, W., Maas, R. E., and Takahashi, S.
    (1987). *Applications of computer technology to
    pre-college calculus: First annual report*, Techni-
    cal Report 310, Institute for Mathematical Stud-
    ies in the Social Sciences, Stanford University.

18. Symbolics, Inc. (1985). *Vax Unix Macsyma Ref-
    erence Manual: Version 11.* Symbolics, Inc.

19. van Vliet, J. C., (ed.) (1988). *Proceedings of the
    International Conference on Electronic Publish-
    ing, Document Manipulation, and Typography*,
    The Cambridge Series on Electronic Publishing,
    Cambridge, England, 20–22 April 1988. Cam-
    bridge University Press. Nice, France.

20. Wolfram, S. (1990). *Mathematica: A System
    for Doing Mathematics by Computer*. Addison-
    Wesley, Redwood City, California, second edi-
    tion.

A Learning Theory Approach to Calculus *

Ed Dubinsky
Purdue University

1 Introduction

The prime question with which we are all concerned is, What can we do to help students learn calculus? Many people are trying to find and/or implement answers to this question. Surely our answers must be concerned both with how we think students learn mathematics and with how we view the content of the field called calculus.

Much of the work in calculus reform has emphasized the content. How can we get the material closer to applications? Which topics can be omitted and which can be added because of the existence of computers and various kinds of mathematical software? Is it good or bad that the content of calculus today is not much different from what it has been for several decades? These are important issues and their resolution will have and should have an effect on what happens to calculus.

On the other hand, I think there is a more important consideration. In a sense, considerations of content are mainly about the mathematician, the teacher, and not about the student. Put more bluntly, I submit that if our goal is for students to learn calculus, then, what is taught and how it is taught is of no importance whatsoever. In the last analysis, the only thing that really matters is what is learned and how that happened—or didn't happen. For this reason, my contribution to this volume will place a major emphasis on how people learn mathematics.

There is another reason for being explicit about our ideas of how people learn. Most mathematicians are thoughtful about their teaching and make many choices about what to do and what not to do in their classes. I suggest that each choice that a teacher makes is determined (usually implicitly) by her or his beliefs, both about how people learn as well as what is the content of calculus. Indeed, I indicate in the next section how I think these beliefs can lead to particular teaching choices. If I am right and this is really what happens, then it seems to me that we should look carefully at these beliefs and decide explicitly whether we really accept them, especially in light of the teaching choices they lead to and the alarming results of these particular choices.

I have looked at my own beliefs about learning. Indeed, I have tried to influence them by doing research and looking at the research of others. As a result of this work, I am beginning to develop a theoretical perspective of how people can learn mathematical concepts and I outline this perspective in Section 3.

In Section 4, I discuss pedagogical implications and lay out some of the choices about teaching that seem to follow from the beliefs about learning to which my theory has led me. In particular, I have come to certain conclusions about how computers can be used, and this has led me to use them in ways that may be somewhat different from what is generally done these days. For example, although computer algebra systems are important and useful, I think that there are alternative, equally powerful ways to exploit computer technology, and I would call for rather more of a balance than exists today. I definitely think that programming, as opposed to using packages or systems, should play a prominent role in the things students do in order to learn mathematics. I hope that these points of view do not set me too much apart from the general thrust of this volume. In any case, I will try to present arguments that support them, especially in Section 4.3.

*This paper is based on a talk given at the St. Olaf Conference on Symbolic Computing Systems, Oct. 20–22, 1989.

2   On Beliefs and Choices

As teachers, we all make choices about what we do in a classroom. Whether we are aware of it or not, these choices are to some extent determined by the beliefs that are part of our conception of how people learn mathematics and what mathematics is.

How people learn: Here are four possible beliefs that one might hold about how people learn.

1. *Spontaneously.* If you believe that students learn mathematics individually and spontaneously by looking at diagrams or listening to a speaker, and that little can be done to help directly, then your answer to the prime question of what we can do to help students learn calculus might be to present material to them in verbal, written, or pictorial form, and expect them to learn it on their own.

2. *Inductively.* If you believe that students learn inductively by working with many examples, extracting common features and important ideas from these experiences, and organizing that information in their minds, then your answer to the prime question might be to have your students spend a very high proportion of their time with examples.

3. *Constructively.* If you believe that students learn by making mental constructions to deal with mathematical phenomena, then your answer to the question might involve a study of just what these constructions are, how they can be made, and what can be done to induce students to make them.

4. *Pragmatically.* If you believe that students learn mathematics as a response to problems in other fields, then your answer might involve introducing students to many applications.

The Content of Calculus: Correspondingly, there are four categories of beliefs about the nature of calculus.

1. *Knowledge.* If you believe that calculus is a body of knowledge that has been discovered by our society (over several hundred years) and that we must pass it on to future generations by transferring it from our minds to the minds of our students, then you might present the mathematics to the students who must somehow imbibe it.

2. *Techniques.* If you believe that calculus is a set of techniques for solving standard problems, then you might have your students spend most of their time practicing these techniques on large collections of problems.

3. *Thought.* If you believe that calculus is a set of ideas that individual and collective thought has created, then your teaching goal might be to help students construct these ideas on their own, with however, a great deal of guidance that will allow them to "stand on the shoulders of giants."

4. *Applications.* If you believe that the essence of calculus is its power to describe, explain, and predict phenomena in the physical world, then your course might be about topics in the physical sciences with emphasis on the role of mathematics.

It is possible, therefore, to make some inferences about what we believe by looking at what we do when we are teaching. I think it is fair to say that the overwhelming majority of calculus teaching is based on the belief that mathematics is learned spontaneously and inductively, and that calculus is some combination of knowledge and techniques. My feeling, after 30 years of teaching calculus and talking about it with other people is that most mathematicians **say** that they believe people learn from examples (inductively) and that calculus is a body of knowledge we have discovered, but they **teach** as if they believe that students learn spontaneously by listening and watching and as if what they are supposed to learn in calculus is a collection of techniques for solving standard problems.

I think this is all wrong. I think that at least part of the blame for the failure of calculus teaching and learning in this country (and perhaps elsewhere) must be placed on the teaching choices we have made. These choices are not wrong because they fail to follow logically from reasonable assumptions (beliefs). They **do** follow and the assumptions **are** reasonable (as I have learned from a multitude of commons room discussions.) But reasonable is not enough. If we want to explain the disastrous results of our calculus courses, the high attrition and failure rate, the low levels of understanding students bring to the science, management and engineering courses which rely on calculus, and the apparent turning away from mathematics by the brightest and best of our students; if we want to significantly improve calculus learning, I think that it is time to question the original assumptions, the beliefs about learning and content, and to ask if other beliefs and instructional treatments they imply might be more appropriate and effective in reaching the ultimate goal.

Indeed, we must do more than question. Perhaps the issue of the content of calculus can be resolved by conferences, panels, special sessions and *ad hoc*, isolated projects. The question of how people learn cannot. For this, research is needed. I don't mean the sort of mindless statistical analyses that overflow the journals in some fields. I mean hard thinking about theoretical notions, and I mean teaching experiments that simultaneously follow from these notions and which are designed to enhance and sharpen the theory.

For the past several years I have been engaged in an attempt to produce that kind of research. My results have led me to the belief that mathematics is more a manner of thought than just knowledge and techniques—that people learn mathematics constructively rather than spontaneously or inductively. It has also led me to wonder about the role of applications, for which there are both positive and negative things to be said.

In the next section I will try to argue for my constructivist (in the epistemological not mathematical sense) beliefs, after which I will present a corresponding theoretical analysis of the learning process along with its implications for teaching and the use of computers.

3 The Learning Process

3.1 Why Constructivism

Let me begin by saying why I tend to reject the idea that mathematics is mainly learned spontaneously or inductively. One argument is, as I have already pointed out, that this belief is what got us to where we are now, so it is at least suspect. Another is the passive nature of learning spontaneously or inductively. We all believe that one has to be active to learn mathematics, but there is nothing active about receiving presentations. Working through examples is certainly an activity, but what happens at best is that the student is active about learning the specifics of the examples. Taking note of common features and organizing material may or may not happen and, in any case, such cognitive activities do not seem to be sufficiently powerful for learning mathematics.

The question of active versus passive in learning mathematics relates directly to the idea of symbolic manipulation. For me this terminology emphasizes the use of tools (equation solver, differentiater, integrater, grapher, etc.) to look at examples, investigate the properties of phenomena, and run through the solutions of problems. All of this is important, but so is the construction of the tools themselves; symbolic computer systems do not pay enough attention to the nature of these tools. To use a popular analogy, I agree that it is not necessary to understand how an engine is made in order to drive a car, but in taking long trips, I have always felt more comfortable when there was someone present who knew something about how the insides of the car work. When a problem arises, I would at least like to know enough to decide if I can deal with it myself, or whether a specialist should be called.

Constructing mathematical concepts rather than merely using them—that is, doing as opposed to observing—is what I was referring to in the Introduction when I talked about alternatives to computer algebra systems. Symbolic computer systems are about how you can **use** mathematics. I believe there are other tools, such as mathematical programming languages, which can help students learn something of what mathematics **is**. I have been trying to find or develop a theory of learning mathematics that makes sense to me as a mathematician, that explains some of the phenomena I have observed, and can guide the development of teaching methods that involve students more actively in the use of computers.

In trying to relate various learning theories to reflections on my own experiences as a mathematician, I have been most impressed by the constructivist theories of Jean Piaget [3]. Mathematical knowledge is not so much something that you **have**, but rather something that you **do**. The observations and experiments I have conducted lead me to believe that what you do is to construct things, to construct them in your mind.

3.2 A Theoretical Approach to Learning Mathematics

3.2.1 What is Mathematics?

One cannot think about how mathematics is learned unless one has some idea of what mathematics is. I would like to explore this epistemological question, beginning with the idea of mental constructions.

Mental constructions are not made for the purpose of maintaining a storage closet of mathematical entities, but for solving problems. I think of "problem" in a very general sense, including applications and also the problem of understanding a certain situation. Most important to the construction of mathematical knowledge is the nature of the problem situation that drives the construction.

All of us who teach are familiar with the distinction between the problem that the teacher sets before the student and the student's perception of that problem. We have every opportunity to make the situation as clear as we can, but eventually we reach the point where the student decides what he or she thinks the problem is and it is that perception that determines the construction. We might think, for example, that the "problem" is to understand that the derivative of a particular function is a useful linear approximation to it near a point. Our students, however, will often conclude that the "problem" is to find a formula for the derivative of that function. It is not very easy to get them to revise their interpretations of the situation in which they find themselves.

Another familiar experience is the inconsistency of mathematical understanding within a single person. Like the athlete, the mathematics student is sometimes able to function at a very high level of sophistication, but on other days, with similar or even identical problems, he or she is much less successful. Thus mathematical understanding is not about what a person is surely able to accomplish, but what he or she has a tendency to do.

In connection with this, we can think of a person, faced with a problem situation, not as bringing forth immutable pieces of mathematical knowledge, but rather as reconstructing what he or she has previously constructed in order to deal with the present situation. It is this reconstruction that leads to inconsistency. Sometimes it produces tools that are less powerful than the person has previously used. On other occasions, stimulated by the special difficulty of the actual problem, it can produce something that is more powerful, more sophisticated, and more effective. In this case, we can say that growth of the person's mathematical knowledge has taken place.

I have tried to put all of these thoughts into a single, compact definition.

> A person's mathematical knowledge is her or his tendency to respond to certain kinds of perceived problem situations by constructing, reconstructing, and organizing mental processes and objects to use in dealing with the situations.

In order to make use of such a definition, it is necessary to be more specific about what is meant by mathematical processes and mathematical objects. Even more important, however, is to understand something about how processes and objects are constructed, since this is precisely what we want to help our students do. My theory provides responses to these questions. They are the result of a combination of analysis and experimentation always flavored by my own understanding of the mathematics involved. Reports on the research that led to what I am about to describe can be found in the papers listed in the references [2,4,5,6].

3.2.2 Construction of Objects and Processes in Mathematics

Roughly speaking, processes are built up out of actions on objects and ultimately converted into new objects which are used for new processes and so on, as a person's mathematical knowledge spirals up to higher and higher levels of sophistication. For the purpose of discussion we need to break into this circular spiral at some arbitrary point.

Numbers are **objects**. **Actions** can be performed on these objects, for example by making arithmetic calculations. When an action such as adding three to a number is repeated with different numbers, there is a tendency to become aware of and **interiorize** this action into a **process**, $x + 3$. This leads to algebra. Processes can also be constructed by composing two processes, say adding three and squaring which gives $(x + 3)^2$ or by reversing a given process, say adding three, to obtain the process of subtracting three or $x - 3$. A single process such as adding 3 can be **encapsulated** to become an object, in this case, the expression $x + 3$. Now the standard algebraic manipulations with expressions can be seen as actions on these new objects. Figure 1 below displays these different kinds of constructions schematically.

There is an interesting observation that can be made from this analysis. The mathematical notation $(x + 3)^2$ can represent two kinds of mental constructions. One is the process obtained by composing the "add 3" process with the "square" process. The other is the object obtained by squaring the object $x + 3$. I think that although both representations are important, the former carries much greater mathematical

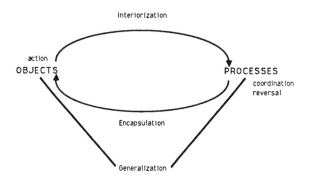

Figure 1: Construction of objects and processes

content, whereas the latter is more formal and can lead to performing manipulations without much understanding of their content. One of problems I have with computer algebra systems is that they emphasize the latter view of expressions like $(x+3)^2$ and do little to encourage the construction of the processes which, in my opinion, carry most of the mathematical content.

There are, of course, much more mathematically sophisticated situations. In calculus, most of the examples in which mathematical concepts are interpreted in terms of objects and processes have to do with functions. For instance, the input/output point of view treats a function as a process, as does the interpretation that points on a graph come from evaluating the function at the x-coordinate and taking the answer for the y-coordinate. On the other hand, considering differentiation as an operation that transforms one function into another, thinking about iterating the composition of a function with itself, or seeing that the solution of a differential equation must be a whole function (rather than a number or even a structured collection of numbers) requires the interpretation of a function as an object.

Sometimes it is necessary to conceptualize a function simultaneously as a process and as an object. To understand the notation $f'(3)$, for example, requires the idea of transforming a function to its derivative (object) and evaluating at 3 (process). A principle of my theory is that this can only happen if the object conception came as a result of encapsulating the

corresponding process conception. In this case, one is able to go back and forth between the two interpretations.

Functions are not the only context in which this theory operates. Mathematical induction is an important tool in calculus, and one usually assumes (wrongly) that it is well understood by the students. Predicate calculus is critical for understanding the $\epsilon - \delta$ formulation of the limit, and I feel strongly that the difficulty students have with limits and related concepts is due to their lack of understanding of quantification. In several papers, I and others have analyzed these concepts in terms of the theory discussed here and similar theories ([7,8,9,11,12]).

The scheme in Figure 1 illustrates the essential ingredients of the constructions in this theory of learning. It can be used to describe a number of mathematical concepts at all levels of sophistication. Indeed, one of my working assumptions is that the construction of **every** mathematical concept can be analyzed in this way. I will point out later how these analyses can be used to guide the design of instruction.

One final point regarding this theory is the notion of **reflection**. An important aspect of learning mathematics is to be aware of the mathematical operations that you perform—for example, when making calculations—and to reflect on their meaning. This is a critical point for the constructivist theory. It enters explicitly in that both interiorization and encapsulation arise, at least in part, as a result of reflecting on the problem situation and methods of dealing with the problem—both successful and unsuccessful.

3.3 The Role of Applications

My thoughts about applications are at least as much at variance with "conventional wisdom" as the other things I have talked about in this paper. They are not, however, nearly as considered—that is, based on research efforts—therefore, what I say will be very brief and should be taken only as raising certain questions, not as providing much in the way of answers.

Why are applications important, anyway? Most of the historical development of mathematics has been driven by a desire to apply it to problems in the physical sciences. In this century, that has been expanded to include some of the social sciences and computer science. Many people conclude from this that applications in a calculus course will make the material more exciting and more relevant.

I am not completely convinced. At the very least

there is the danger that a mathematics course with too great an applications component could evolve into a course about the applications field. We must keep in mind that our role is to help students learn mathematics primarily and not physics or chemistry or management.

I think that in looking into the role of applications, we must think mainly about what will motivate a student to reconstruct her or his mathematical knowledge on a more sophisticated level. A part of my theory that has not been emphasized here suggests that the need to solve a problem, to straighten out confusion, to deal with a previously unmanageable situation is a basic human drive, on a par with the need for food or sex. If this is the case, then teachers can think of harnessing this drive and using it to provide students with energy to work hard, to overcome the confusion and frustration that is an endemic feature of learning mathematics.

This is how applications should be used, and I see no reason why applications to the physical world, or to any subject outside of mathematics, is necessarily the only context for exploiting this basic human drive. Applications within mathematics can work just as well for some people. I see equal value in interpreting the usual method of finding extreme points as an application of the theory of solving $f(x) = 0$ as I see in interpreting the integral of a certain function as work.

The key, in my opinion, is the value of applications in setting up problem situations for students that will motivate them to construct mathematical concepts. In doing this, we must not only look at that which intrigues us, but we must also study how various applications will appear to students and what mental constructions will result from students working with them.

4   Pedagogical Implications

4.1   Teaching

My choice of beliefs about what mathematics is and how it can be learned has affected my selection and design of instructional treatments of mathematical topics in a number of ways. The most important of these is that the learning theory I outlined in the previous section implies that the primary goal of teaching is to help students construct appropriate processes and objects and get them to reflect on and use these constructions in dealing with problem situations that

arise in mathematics. This means that each topic must be analyzed to determine just what constructions one would like the students to make, and then it is necessary to design problem situations that will induce them to make these constructions.

That is a fairly complete description of my research and development program in learning mathematics. It involves detailed observation of students, experimentation to determine the effect of various instructional treatments on what students do or do not construct and, throughout, it makes full use of my own personal understanding of the mathematics that I would like students to learn. This program has been in operation for about six years and I hope it will continue for a long time to come. Results, at this early stage, are of necessity incomplete, but there are a few areas in which there seems to have been some progress and these have been described in the literature. My purpose here is not to discuss this research program at length, but rather to rely on it to talk about helping students learn calculus. In particular, I would like to consider a few examples of topics in calculus, describe how they can be analyzed in terms of this theory and point out some interesting (and often misleading) ideas students may develop as their first reactions to them. Then I would like to describe some instructional treatments using computer systems that, in my experience, have made quite a bit of difference.

Before doing that, I would like to point out that this particular research approach is not the only way of improving learning in mathematics. A number of individuals throughout the world are experimenting with innovative pedagogical approaches. These range from small-group problem solving to the use of writing in mathematics. A particularly interesting approach is that of M. Legrand and D. Alibert in Grenoble, France. They work with large classes (about 200) in introductory analysis and claim to be able to conduct "scientific debate" in those classes [1]. *UME Trends* runs a regular column on teaching innovations, and the MAA Committee on Teaching Undergraduate Mathematics is preparing a book that will describe some of these approaches.

4.2   Some Calculus Concepts

4.2.1   Graphs and Functions

There is a general feeling that graphs are extremely important as visual representations of functions and computer algebra systems are a powerful vehicle for getting students to work with graphs in sophisticated situations. There is some evidence, however, that students do not necessarily make the connection between a graph and the function it represents.

In a recent paper, Paul Goldenberg [10] gave examples in which computer representations of graphs seemed to lead to misconceptions on the part of young children. I have seen undergraduates in many situations apparently fail to make the desired connection.

In a continuing survey I have been making for several years, I ask undergraduates (many of whom have worked with computer algebra systems) two questions: What is a function? and Give one (or two or three) examples of a function. I have collected hundreds of examples of responses to these questions and I am quite surprised that less than 10% of them contain any reference whatsoever to a graph.

Another example has to do with functions defined in parts (i.e., using different expressions for different parts of the domain). Students often do not consider such an example to be a function or they consider it to be several functions instead of one [13].

Here is a recent experience that surprised me. I asked a class of 30 calculus students about the limiting position of the secant. I drew the standard picture of a curve on the blackboard with coordinate axes and two points on the graph connected by the secant. Then I moved one point closer to the other, redrew the secant and repeated the process one more time. I asked the class what they thought would happen to the line I was drawing as the second point continued to move towards the first. About 10 of them said it would be the tangent at the fixed point. But another 10 said it would be the horizontal line through the fixed point and 5 others said it would be the vertical line through that point. The remaining 5 gave "miscellaneous" responses. The 15 horizontal and vertical line students stuck to their guns strongly. When pushed for explanations they said that the vertical (or horizontal) segment was going to 0 so the line would become horizontal (vertical). When confronted with the fact that the other segment was also going to 0, they said things like, "Oh, I didn't pay any attention to that."

This experience suggests to me that the tangent as limiting position of the secant may not be an intuitive notion for students and they may not be connecting the pictorial situation with the corresponding difference quotient.

In terms of my theory, the conception of function that is important here is the process conception in which the function transforms values in the domain to values in the range. A graph, however, is an object. If it does not arise through encapsulating the function process (for example, if it arises for the student by looking at pictures drawn on a blackboard or by entering expressions to a computer algebra system and pressing a button) then, I suggest, the connection with a process may not be constructed in the student's mind and this could explain why students may not be developing a reasonable intuition about the derivative being the limiting slope of the secant. It might also explain some of the other difficulties we have seen including those mentioned by Goldenberg [10].

4.2.2 The Fundamental Theorem of Calculus

One can start this circle of ideas with the derivative, beginning with the process of using the difference quotient to estimate the derivative at a point and passing to the limit, encapsulating that process to an object which is the derivative at a point and then varying the point to obtain a function. This entire process is interiorized to obtain the notion of taking a function f and transforming it to another function f'. Conceptually, this requires that the student be able to think of a function as an object to which an action can be applied, resulting in a new object. This is a higher level action which again must be interiorized to obtain a process. In this way, the concept of function is generalized to include functions whose domains and ranges are sets of functions.

The integral is analogous. First, there is the process of partitioning an interval and measuring rectangles to estimate an area. Passing to a limit is more complex here than in the case of derivatives. Nevertheless one can hope that the student will construct a reasonable mental process that, given a function and an interval, produces a number. This process must then be encapsulated so that one endpoint of the interval can be varied so as to obtain a new function whose value at a given point is the encapsulated process (of computing the integral). Thus, again, one interiorizes a process that acts on functions to produce new functions.

The last step in the entire scheme is to coordinate the two processes by composing them. One can apply the integration process and follow it by the differentiation process, and altogether one has again a single process. It seems important for students to construct this process first and then see that, when applied to reasonable functions, it gives back the original function. At this point, one can compose the two processes again, this time in the opposite order to investigate the complication that arises from the fact that the integration process is not one-to-one.

The cognitive difficulties for students in trying to make the mental constructions described here are immense, and the fact that traditional methods of teaching do little to help them directly in this endeavor goes a long way, in my opinion, to explain why so few of our students emerge (if they do emerge) from the calculus course with any appreciation for, much less understanding of, the incredible intellectual achievement that this theorem represents.

4.2.3 Sequences and Series

Considering sequences and series only briefly, we note that, once again, we have the function concept, this time in the context of functions whose domains are the set of positive integers. Such a function is a sequence and the process conception of function is a critical ingredient in understanding, for example, properties such as monotonicity, bounded, convergence, constant, alternating, and so on.

The sum of the terms of a finite sequence is also an important process; its encapsulation is necessary before one can consider, relative to a given sequence, a new function which gives, for each index, the sum of the terms of the sequence up to that point. According to my theory, this would be one reasonable way to come to the notion of sequence of partial sums, from which one can then go on to discussion of various topics related to infinite series.

4.3 Use of Computers

Finally, after a philosophical discussion of the relationship between teaching methods and beliefs about learning, a description of my own theoretical approach, and an indication of the kinds of analyses of mathematical concepts that I feel must precede the design of specific instructional treatments, I am ready to say something about the use of computer systems in implementing these designs.

I can begin with a straightforward point about the issue of having students use prepared software packages versus having them program in some general purpose language. I believe that a conclusion follows from all that I have said so far and it goes like this. Using a software package that performs some mathematical operations is another way of **showing** the mathemat-

ics to the user. Writing computer code that implements a mathematical process or represents a mathematical object is a **construction** by the programmer. Moreover, my experience has convinced me that anytime you construct something on a computer then, willy-nilly, you are constructing something in your mind. Since, according to my theory, learning mathematics consists of constructing certain processes and objects, it follows for me that the use of computers should emphasize programming and that programming tasks for students should be designed to influence them to make constructions that will contribute to the growth of their mathematical knowledge.

Two caveats. First, I am saying only that programming is more important than using software packages, not that the former should be employed to the exclusion of the latter. Indeed, I will show below an example of coordinating the two kinds of computer systems and how this can lead to particularly useful tasks for students.

The second caveat is that my entire position collapses on practical grounds unless the programming language and the environment in which it is used are extremely convenient and relatively free of frustrating syntax issues that are not connected with the mathematical issues. (Some of the difficulties are connected with the mathematical issues, and in that case the student's frustrations are the right difficulties to struggle with!)

I hope that the examples given below suggest that we are at least trying to avoid these two pitfalls.

4.3.1 *Maple*, ISETL, and your Favorite 3D Graphics System

These are the systems we use. I believe that most readers of this volume will be familiar with *Maple* and presently available 3D graphers that don't do much more than display a picture when an expression is entered. In both cases, there are many similar systems that could be used to do the things our project is doing and the materials being produced could easily be adapted to any of them.

The same is true, in principle, of *ISETL*. At the moment it is the only language I know of that can be used conveniently in the way I will describe. I am sure that other, better systems of this kind will arise and, again, I expect that our material will be easily convertible to them.

In the meantime, I feel that not only is *ISETL* unique today, but that few people are familiar with it and so I must explain some things about its principles and features before discussing how all of these sytems are used.

ISETL is an interactive, interpreted programming language that implements a number of mathematical constructs in a syntax that is very similar to standard mathematical notation. This is the reason why using the language does not violate my second caveat.

One uses *ISETL* by entering an expression to which the system responds by evaluating and returning a result. An expression can involve arithmetic operations on numbers (integers or floating point), boolean operations, and operations on character strings. Assignments can be made to variables, and expressions can combine variables and constants. The domain of a variable is determined in context dynamically (it can change) by the system and there is no need to declare data types, sizes, etc. Many important mathematical operations on these data types are implemented directly in *ISETL* and are used with a single command. In addition to the usual arithmetic, they include mod, max/min, even/odd, signum, absolute value, random, greatest integer less than, concatenation (of strings), and the standard trigonometric, exponential, and logarithmic functions.

The power of *ISETL* begins to appear with the complex data types of set, tuple, func, and map.

Syntax such as

```
{7..23};
{-4,-1..40};
{9,7..0};
```

can be used to construct sets of finite arithmetic progressions of integers. It is also possible to construct a set containing any data types whatsoever (including other sets) simply by listing them. For example, the following set has cardinality five.

```
{8-1, "t" + "he", 1.2, {1,3,4,2},
    { {1..4}, 3<2, "the", 7, false}};
```

Once such sets have been constructed, one can then construct complicated subsets by using a set former notation that can generally be understood by anyone who knows the mathematics. For example, here is a set of cubes of even integers (of absolute value less than or equal to N) whose squares are congruent to 2 mod 4.

```
{k**3 :  k in {-N,-N+2..N} | k**2 mod 4=2};
```

Standard set operations are implemented with single command syntax. They include union, intersection, difference, adjunction, tests for membership or subset, power set, cardinality, and selection of an arbitrary element. It is possible to iterate over a set to make loops conveniently, but the operations of existential and universal quantification over a set are implemented and they often render loops superfluous. For example, in the following code, the first and second lines construct, respectively, the sets of all positive even integers and all primes less than or equal to N and the third checks the Goldbach conjecture up to N. The last line returns the value **true**.

```
E := {2,4..N};
P := {p :  p in {2..N} | (forall q in
    {2..p-1} | p mod q /= 0 };
forall n in E | (exists p,q in P | n = p+q);
```

The *ISETL* tuple is a finite sequence. It is the same as an array in standard programming language usage and may be considered as an infinite sequence, only finitely many of whose terms have been defined. One can change a single component of a tuple (and possibly increase the number of components that have been defined), concatenate two tuples, or add a term to the end. If any of these operations affect the length of the tuple, the change is made automatically. The same iteration and quantification operations can be performed over tuples as well as sets. The syntax for tuple is very similar to that for sets except that square brackets [ ] are used instead of curly brackets.

The name for procedure in *ISETL* is func. A func is a function; it accepts parameters and returns a value. Here is an example of a func that represents a function that is defined by different expressions on different parts of its domain. It is continuous but not differentiable at $x = -1$, and at $x = 1$ it is once, but not twice differentiable.

```
f := func(x);
      if x < -1 then return -x**3
      elseif x < 1 then return x**2;
      else return 2*x-1;
      end;
    end;
```

A func may be assigned to a variable (such as f in the above example) and then standard evaluations such as $f(4)$ have the usual meaning. The func is the main tool used to help students construct a process conception of function.

It is also possible to implement a function in *ISETL* as a set of tuples each of which has length 2, that is as a set of ordered pairs. This feature is used to work with functions given only by a table of data.

Funcs are data and can be treated as such. They can be passed to procedures as parameters, constructed by a procedure and returned, or operated on as with ordinary data. This feature is critical in helping students construct a concept of function as object.

The main evidence I can offer to convince you that my second caveat is avoided is to point out that I and others have used *ISETL* in mathematics courses many times over the last four years. At the very beginning the students spend about 90% of their time with programming issues (some of which are not unrelated to mathematical concepts) and the rest of their time thinking about mathematics. This evolves over a period of about two weeks after which time it is reversed and student thinking is at least 90% about mathematical ideas (although sometimes they are in the guise of a programming issue, such as the task of writing a func which returns as its result, not a number but a function).

4.3.2   Examples of Concepts and Computers

Now I will try to give some indication of how these computer systems are used to help students make the mental constructions that, according to the theory described in this paper, will lead them to understand mathematical concepts. I will look at the three examples considered in the previous section and try to relate the computer use to the epistemological analysis that is sketched there.

Graphs and Functions: Our instructional approach considers that the connection between the production of data points and a curve on a coordinate system is of paramount importance. Here is what we do to help students construct this connection in their minds.

They are given the task of writing *ISETL* funcs to represent various functions (such as `f` in Section 4.3.1). Then they can write a line of code such as

```
graph(f,a,b,n,name);
```

Here `f` is the variable they have used to store their representation of the function, `a`, `b` the endpoints of an interval and `n` is the number of points (evenly spaced) on this interval at which they wish to sample the values of `f`. The student must choose all of these to determine which function is to be graphed, on what interval and how finely. The result of this command is to place in the file **name** a table of values of `f` from `a` to `b` at `n` points.

The file can be viewed but this is not required. The student then leaves *ISETL*, enters *Maple* and performs the following two commands.

```
read name;
sketch ";
```

The result is that the table appears on the screen and then the points are plotted on a graph.

One might question the use of such a multi-step operation. My feeling is that not only does it give the student an opportunity (encouraged by the teacher) to reflect on what is going on, but it is closer to the way things work in scientific investigations than the one-shot, push a button approach to which some people are reducing these powerful computer systems.

In any case, the student can study the behavior of the function at the two interesting points, $x = -1, 1$. The corner at -1 will be obvious, but one does not see anything at 1. To study this, the student can write a function that will approximate the derivative of f. A func for this would look like

```
fp := func(x);
      if x < -1 then return -3*x**2
      elseif x < 1 then return 2*x;
      else return 2;
      end;
   end;
```

and the same process can be applied to `fp`. It is possible to use our scheme to place `f` and `fp` on the same graph.

We have students do all this with a number of functions before getting them to write the following *ISETL* program.

```
ad := func(f);
   return func(x);
      return (f(x+0.00001) - f(x))/0.00001;
   end;
end;
```

This presents them with some very serious cognitive difficulties related to an object conception of function, but the reward is powerful for the student who eventually learns to write `ad(f)` or `ad(f)(3.7)` and understand the meaning of the computer's response.

Learning is further enhanced by the opportunity to vary the function f to which `ad` is applied. This gives an added dimension to the general technique of having students study a picture on which appears both a function and its derivative without indication of which is which. I believe that when these pictures come from functions that the student constructed on the computer the whole experience becomes richer.

The fundamental theorem of calculus: Working with the above func **ad** helps students take the cognitive steps described in Section 4.2.2 and construct a notion of the derivative as an operator that takes a function and transforms it into another function.

The next step in the circle of ideas presented in Section 4.2.2 is to construct the concept of definite integral. The key computer activity for the students here is to construct a number of *ISETL* funcs that implement various Riemann sums. Here is one that does it by always taking the left endpoint.

```
RiemLeft := func(f,a,b,n);
    x := [a+((b-a)/n)*(i-1) :
            i in [1..n+1]];
    return %+[f(x(i))*(x(i+1)-x(i)) :
            i in [1..n]];
    end;
```

Notice that, except for using % in place of $\sum$, the syntax here is correct mathematical notation. Future enhancements of the language will probably use things like $\sum$. There are variations corresponding to taking right endpoints, midpoints, minimum and maximum. The trapezoid rule is included in the midpoint variation and Simpson's rule is very easy to implement. The students apply these funcs to many functions, draw pictures on graphs produced by *Maple* and see that the mathematical calculation does approximate the area under a curve, or between two curves, for example. The standard properties of the integral (linearity, monotonicity, concatenation of intervals) are more easily understood by the student because they correspond to properties of computer constructs that he or she has made. Even limits here can begin to make sense for the student.

The students are helped to encapsulate the process of computing a Riemann sum to obtain an object that is an approximation to the indefinite integral, by writing the following func that corresponds to the step of varying the upper endpoint to obtain a function.

```
Int := func(f,a,b);
    return func(x);
            if a<=x and x<=b then
                return RiemLeft(f,a,x,25);
            end;
    end;
```

Students have a hard time writing this. In working with them it seems to me that their struggles are focused on the non-trivial mathematical concepts of using the definite integral with varying endpoint to define a function.

But the "crunch" really comes when the students are asked to write code that will produce tables of values of functions obtained by differentiating first and then integrating, integrating first and then differentiating, and just evaluating the original function. They had earlier been given a func called **print1** which they can adjust to print various tables. Now they receive the following instructions.

> Adjust the func **print1** to obtain the func **print3** which will accept a func representing a function f, an interval $[a,b]$, a positive integer n and a file name. The result of **print3** is to place in the file four columns of values. The first gives the numbers x at $n+1$ evenly spaced points in the interval $[a,b]$; the second gives the value of f at x; the third gives the value obtained by applying **ad** to f, applying **Int** to the resulting func and then evaluating the resulting func at x; and the last column gives the value obtained by applying **Int** to f, applying **ad** to the resulting func and then evaluating the resulting func at x.

Adjusting the func is a minor technical task which gives no trouble. The really difficult part is that they must work out for themselves the following expressions in *ISETL*.

```
Int(ad(f),a,b)(x);
```

```
ad(Int(f,a,b))(x);
```

and this is a major difficulty for them. Many students overcome the difficulty and all of them seem to get something out of the experience. The file that they produce can then be given to *Maple* with which they can produce three graphs for each function. Two of them are almost identical and the third is displaced by a constant amount. At the end, most students figure out for themselves that the displacement corresponds to the value of $f(a)$ and it is the "constant of integration" that they have seen before.

Sequences and series: All that I would like to say about sequences at this point is that the students can write funcs such as

```
s := func(n);
        return (-1)**n/(n*(2*n+1));
    end;
```

and use them to study the behavior of a sequence. They can also write a func which converts a sequence to the corresponding sequence of partial sums

```
sums := func(s);
    return func(n);
        return %+[s(i) :  i in [1..n]];
        end;
    end;
```

and then work with expressions such as `sums(s)(14);`.

All of this is designed to help them take the mental steps mentioned in Section 4.2.3.

5   Conclusion

I have presented a very small number of examples in this paper. We have designed and implemented at Purdue a first version of a three-semester (total of 14 credit hours) calculus course based on the learning theory presented here. The course involves a Macintosh lab in which students use the computer in ways that I have described above. In class, we replace lecturing with small group problem solving. The computer activity and the problem-solving work are all designed to help students make the mental constructions we feel are necessary to learn the mathematical concepts in calculus.

Our first version of the course stayed very close (in content and sequencing, but not in pedagogy) to the standard syllabus used at Purdue. We are in the process of implementing a second version in which our teaching methodology was developed and the material was completely revised. The revision of content is based on the learning theory and makes use of the needs of departments whose students are taking calculus as a service course. It exploits the computer to deemphasize certain topics and include others. We are beginning to conduct experiments to find out more about how students learn some of the mathematical concepts that appear in calculus. This will also contribute to the design.

Our overall, long-term program consists of: coordinating a learning theory with experimentation to analyze each topic in calculus from an epistemological point of view; determining mental constructions that students can make in order to acquire the concepts

necessary to learn the mathematics in the topic; designing computer tasks and problem situations that will induce them to make these constructions; and synthesizing all of this into a practical system that can function in the "real world" of an academic course without too much extra cost or disruption of normal activities. It is a daunting program, and I cannot say that I am optimistic about succeeding to bring it off in its entirety. But I do believe that the problems of the calculus course are profound. No single, simple, short-term program is going to help very much. At the very least I am convinced that the magnitude of our approach is appropriate and necessary—and I hope to at least make a good start on it.

References

1. Alibert, D. (1988). "Codidactic system in the course of mathematics: how to introduce it?" *Proceedings of the 12th Annual Conference of the International Group for the Psychology of Mathematics Education*, (A. Borbàs, ed.) Veszprem, 109–116.

2. Ayres, T., Davis, G., Dubinsky, E., and Lewin, P. (1988). "Computer experiences in learning composition of functions," *Journal for Research in Mathematics Education*, 19, 3, 246–259.

3. Beth, E.W. and Piaget, J. (1965). *Mathematical Epistemology and Psychology* (W. Mays, trans.), Dordrecht: Reidel.

4. Dubinsky, E. (1986). "Teaching mathematical induction I," *The Journal of Mathematical Behavior*, 5, 305–317.

5. Dubinsky, E. (1989). "Teaching mathematical induction II," *The Journal of Mathematical Behavior*, 8, 304 (1989).

6. Dubinsky, E. "On learning quantification," in M. J. Arora (ed.), *Mathematics Education into the 21st Century*, (in press).

7. Dubinsky, E. "Constructive aspects of reflective abstraction in advanced mathematical thinking," in L.P. Steffe (ed.), *Epistemological foundations of mathematical experience*, New York: Springer-Verlag.

8. Dubinsky, E., Elterman, F. and Gong, C. (1988). "The student's construction of quantification," *For the Learning of Mathematics*, 8, 2, 44–51.

9. Dubinsky, E. and Lewin, P. (1986). "Reflective abstraction and mathematics education: the genetic decomposition of induction and compactness," *The Journal of Mathematical Behavior*, 5, 55–92.

10. Goldenberg, P. (1988). "Mathematics, Metaphors, and Human Factors: Mathematical, Technical, and Pedagogical Challenges in the Educational Use of Graphical Representations of Functions," *Journal of Mathematical Behavior*, 7, 135–173.

11. Sfard, A. (1987). "Two conceptions of mathematical notions, operational and structural," *Proceedings of the 11th Annual Conference of the International Group for the Psychology of Mathematics Education*, (A. Borbàs, ed.) Montreal, 162–169.

12. Sfard, A. (1988). "Operational vs. structural method of teaching mathematics—case study," *Proceedings of the 12th Annual Conference of the International Group for the Psychology of Mathematics Education*, (A. Borbàs, ed.) Veszprem, 560–567.

13. Vinner, S. and Dreyfus, T. (1989). "Images and Definitions for the Concept of Function." *Journal for Research in Mathematics Education*, 20, 4, 356–366.

Mathematical Processes and Symbols in the Mind

David Tall
University of Warwick

1 Introduction

Mathematics, as taught to students, is continually the subject of scrutiny to see if it is appropriate for its task. In particular, the calculus, for so long conceived as the essential foundation of college mathematics, is being questioned as to its value in the wider realms of advanced education. Students seem to find much of it so difficult, and now, in the new era of computer and information technology, symbolic manipulators are available which can perform much of the algorithmic work of the calculus, so the question becomes: if a computer can perform the symbolic manipulations, why should we force students to do it? If the computer is available, why not forge a new partnership between student and computer in which each contributes their special abilities to produce a greater whole?

This article will consider the nature of human thinking processes to see how symbolism is utilized in mathematical thinking and to consider how technology is best integrated into the education process. In particular it will look at the kind of thinking a mathematician performs, which seems to make the mathematics so much easier than that faced by the average, or below average, student. We shall see that the child growing into the adult faces problems at every stage, which relate to the divergence between the thinking of successful mathematicians and those who eventually fail. Regrettably, so many of the latter persist into college level mathematics with a way of thinking that makes their method of doing mathematics so much harder for them than the mathematics performed by their professors.

2 The Growth of Mathematical Ideas

The human mind is the product of five million years (and more) of evolution. Yet the growth in mathematical knowledge is exponential with more new ideas being developed each year than have ever occurred before. The foremost Renaissance thinkers could hope to be poets, philosophers, musicians, mathematicians and many other things besides. Today knowledge grows at such a rate that the expert mathematician can no longer hope to encompass the whole of mathematics, gaining expertise instead in a relatively small part of the total.

On the other hand, there is no reason to suppose that there is anything dramatically different about the fundamental human apparatus of thinking than was present say two thousand years ago at the height of Greek mathematics, or ten thousand years ago with prehistoric man. Yet we expect our average student to cope with a knowledge base beyond that available in totality to any previous generation. What is it that enables this growth in knowledge to be encompassed in the minds of ordinary mortals of today's generation? First it is through the use of language, that enables the communication of thought, and through written symbolism that enables the essence of this thought to be passed on from generation to generation. But what is more important still is the manner in which the underlying concepts develop and the way in which the symbolism is used to assist the development of these concepts.

An analysis of the evolution of mathematical ideas shows that different parts of mathematics involve different kinds of thinking processes. Classical Greek geometry arises from observations of properties of specific kinds of objects which are idealized as mathematical models: points, lines, triangles, circles. These properties are described in a general manner which allows constructions to be carried out in a specified way, for instance, drawing the three angle bisectors of a triangle and observing that they are concurrent. Then arises the desire to show that this will always be so, resulting in the concept of geometric proof. The descriptions of geometric objects need to be refocused as definitions that prescribe the mental objects from which deductions can be made. There is a desire to refine the theory to make the definitions minimal (it is not necessary to say that a square has equal sides and four right-angles—with equal sides, one right angle will do). But the symbolism used here: letters for points, two letters for a line, three for a triangle, and so on, all stand for a mental idealisation of objects that exist in reality. The detachment from reality is more a matter of philosophy than fact: demonstrated at the end of the nineteenth century by the realisation that there remained a dependence on geometric actuality because concepts such as "between" or "inside" had yet to be formally defined, but were an implicit part of the theory.

Number and algebra are different. These involve processes which are eventually symbolized in such a way that the symbols act dually for both the process and the resulting concept. This sequence of process becoming concept has become a major focus of mathematics education research in recent years (Beth and Piaget [1]; Greeno [7]; Sfard [11]; Harel and Kaput [8]; Dubinsky [3]; Gray and Tall [6]). It underlies the fundamental growth of modern areas of mathematics: arithmetic, algebra, calculus and analysis. It will play a crucial role in the successful use of symbolic manipulators in education.

3 Symbols Representing Both Process and Concept

Processes are carried out and represented by symbols which subsequently take on a dual role, evoking either the process itself or the product of the process, depending on the context. Thus it is that:

1. 5+3 represents both the process of addition and the concept of sum;

2. 5×3 represents both the process of multiplication (through repeated addition $5 + 5 + 5$) and the concept of product;

3. the symbol 3/4 stands for both the process of division and the concept of fraction;

4. the symbol +4 stands for both the process of "add four" or shift four units along the number line, and the concept of the positive number +4;

5. $3 + 5x$ represents both the process "add 3 to the product of 5 and x" and the concept of the algebraic expression;

6. the function notation $f(x) = x^3 - 27$ simultaneously tells both how to calculate the value of the function for a particular value of x and encapsulates the complete concept of the function for a general value of x;

7. an "infinite" decimal representation $\pi = 3.14159\ldots$ is both a process of approximating π by calculating ever more decimal places and the specific numerical limit of that process;

8. various limit notations, such as:

$$\lim_{x \to a} f(x), \quad \lim_{n \to \infty} \sum_{k=1}^{n} a_k, \quad \lim_{\Delta x \to 0} \sum_{x=a}^{b} f(x)\Delta x, \quad \text{etc.}$$

represent both the process of tending to a limit and the concept of the value of the limit.

What makes mathematical thinking powerful is the flexible way in which this conceptual structure is used. By using the symbolism to evoke a process, it can be used to compute a result, and by thinking of it as an object, it can be used as part of higher level manipulation. This results in a tremendous compressibility of mathematical conceptions. A compact symbolism can represent a complex concept which may also be mentally manipulated as a single entity. This proves to be a powerful tool for the mathematician, though it may cause a barrier for the learner who lacks the flexibility of meaning.

> One finally masters an activity so perfectly that the question of how and why students don't understand them is not asked anymore, cannot be asked anymore and is not even understood anymore as a meaningful and relevant question (Freudenthal ([4], p. 469).

So it is beholden to us as educators, if not as mathematicians, to analyse this process of compressibility to formulate ways in which it might be made available to a wider range of student ability.

4　The Amalgam of Process and Concept as "Procept"

The flexible use of a symbolism as either process or concept, so freely available to the professional mathematician, causes great difficulty for many students. It is well-recognized (e.g., Harel and Kaput [8]; Dubinsky [3]) that the composite of two functions f, g, can be conceived process-wise in the notation $g(f(x))$: first calculate $f(x)$ and then calculate g of the result. But if the composite function $g \circ f$ is to be considered as a mathematical object in its own right, given in terms of the mathematical objects f, g, then a good deal of mental movement from concept to process and back again becomes important.

> Initially, functions are processes and so the subject must have performed an encapsulation in order to consider them as objects. It is important, for example in composition of functions, for the subject to alternate between thinking about the same mathematical entity as a process and as an object (Dubinsky [3]).

The question we should ask ourselves is "do mathematicians consciously think always that they are alternating between thinking of a function as a process and an object?" I think not. Having compressed the ideas (through using symbols), we simply use the symbol to denote whichever mental representation is appropriate, often without realizing consciously what we are doing.

In the minds of successful mathematicians a symbol evokes either process or concept, whichever is appropriate, and this is done so subconsciously that we may be unaware that it is happening. To allow this idea to be a focus of attention, my colleague Eddie Gray and I formulated the term "procept" to mean:

> ... the amalgam of process and concept in which process and product is represented by the same symbolism (Gray and Tall [6]).

This is intended to allow us to focus on the fact that good mathematicians think of a procept in a way that exhibits duality (as process or concept), flexibility (using whichever is appropriate at the time) and ambiguity (not always making it explicit which they are using). The ambiguous use of symbolism is seemingly anathema in mathematical formalism, where definitions are made which quite clearly formulate a concept in one specific form ("a function is a set of ordered pairs such that ..."). Yet, having defined a function as a set of ordered pairs, as a thing, we then blithely go on to use it as a process or as an object, whichever suits us at the time. The question to be asked therefore is: by making this flexibility explicit, can we help students develop these kinds of thinking processes? Before responding to this, let us consider the different kinds of thinking processes that occur in practice.

5　Procedural and Proceptual Thinking

Our research with students of different ages, from kindergarten to college, shows a surprising similarity of difficulty at all levels. In traditional mathematics it is necessary first to acquire the ability to carry out a procedure, and then, after long practice, this is compressed mentally into a more compact mental object, often through the use of appropriate symbolism, to enable the mental object to become the focus of attention at a more abstract level. Students (from an early school age up) initially see the task as conquering the procedure. The more able soon encapsulate the procedure by use of appropriate symbolism and develop a flexibility with the notions that enable them to derive new ideas from old. A child may not know the value of 4×7, but might think of it as "four sevens" and know that "two sevens are fourteen" so "four sevens are fourteen plus fourteen, which is twenty eight." This method of deriving new knowledge from old is a natural consequence of proceptual thinking. I claim that the proceptual thinker has a built-in knowledge generator. It is not necessary for such an individual to work hard to get results; these results are an automatic product of the knowledge structure. I conjecture that this flexible proceptual mode of thought is a major factor in the ability of the more able to do mathematics seemingly with little effort. Such a structure is organic. With a little fertilization it grows naturally almost of its own accord.

The less able, on the other hand, are more likely to focus on the currently required procedure as the main aim of the task. Success for them is being able to carry out the procedure and produce the required answer. Gray [5] observed that children responding to simple arithmetic tasks seek the security of being able to carry out the procedure, rather than the flexibility of being able to derive facts from other known facts. Procedurally oriented children are often quite creative in developing their own methods for carrying out procedures that lead to short-term success. But this can also lead to long-term failure as the personal method may fail to cope with more complex tasks that require encapsulation of the procedure as an object for higher order manipulation.

These early steps in mathematics lead to patterns of thinking that can cause problems in college mathematics. It is my belief, for example, that the difficulties that an average college student has with algebra occur because of previous rule-bound approaches to the subject. When students do not understand what something is, at least they can get temporary success by becoming secure with procedures to do things with it. In the early steps of algebra they meet an algebraic notation which generalizes arithmetic. But whereas the arithmetic symbolism of operations such as addition are linked to a procedure to carry out the process and get the answer, the algebraic symbolism seems more obscure. The symbol $10 - 5x$ represents the process of taking 5 times x from 10, but it is a procedure which cannot be carried out until x is known and, if x is known, why use algebra anyway when arithmetic will suffice? Algebraic symbolism violates many individual's innate understanding of mathematical symbolism which in arithmetic tells them what to do and signals how to do it. The syntax is strange: why is $2 + 3x$ not computed from left to right as $3 + 2$, which is 5, times x ? When students begin to feel uneasy, they often seek security in manipulating symbols to get the right answers. Each new topic is solved by learning a new and often seemingly arbitrary rule, "do multiplication before addition," "do operations in brackets first," "do the same thing to both sides," "cross-multiply," "put over a common denominator," "change sides, change signs," etc. (Tall and Thomas [17]).

We believe that these difficulties with algebra carry through to college students, and that the need for immediate procedural success, if not complemented by meaningful use of notation in the early stages, can so easily lead to meaningless symbol-pushing guided by these arbitrary rules. In Britain the fluency in algebraic manipulation at 16 years old is diminishing, although problem-solving abilities with numbers seem to be improving. The initial introduction of differentiation using the symbolic calculation of limits, even for a simple function like x^3, is severely compromised because it cannot be assumed that the whole population taking the subject can simplify the expression $((x + h)^3 - x^3)/h$. We believe that this will lead to serious problems at college and university which may not be helped by the use of symbolic manipulators, unless this is part of a concerted effort to give proceptual flexibility to the meaning of the symbolism.

At higher levels the same proceptual difficulties recur again and again. Consider, for example, the product of two matrices. The procedural thinker will see the product as a calculation of each entry of the result through looking along a row of the first matrix and down a column of the second matrix, multiplying together corresponding pairs of entries, and adding together the results. The procedural product of two matrices involves a great deal of process. The proceptual thinker will see that this can be represented symbolically as the product AB of matrices A and B and, by thinking of the matrices as single objects and the product AB as an object, can begin to conceive of higher level structures such as $(AB)C = A(BC)$, $A(B + C) = AB + AC$, or that, usually, $AB \neq BA$, and so on. For the procedural thinker, these relationships occur not at the manipulable object level, but at the procedural level, involving far greater detail, far greater cognitive strain, and far greater difficulty for a less powerfully structured mind. No wonder the more able succeed almost trivially, whilst the less able are faced with catastrophic failure.

Once again, if students are procedural in their thinking, then they are faced with greater difficulties than if they develop proceptual flexibility. The same phenomenon occurs in other topics, for instance, in the understanding of limits, where students initially think of $\lim_{n \to \infty} s_n$ as a process of approaching a limiting value. They are faced with new problems here. To calculate the limit of, say

$$\frac{4n + 3}{2n + 1},$$

they may conceive this as "what happens to the calculation when n grows large?" A common suggestion is that "the 3 and the 1 become small, so the answer is roughly $4n/2n$, which is 2." They may develop a genuine intuition which helps them solve problems in a personal manner unrelated to the formal definition, but it may be an idiosyncratic method which fails in another context.

The proceptual structure of the limit definition is quite different from that of, say, arithmetic, where new derived facts may be obtained from old by using the same arithmetic operations. In the formal handling of limit, the student must cope with a difficult definition with several quantifiers. It is an awkward calculation, given an $\epsilon > 0$ to find an N such that

$$n > N \text{ implies } \left| \frac{4n+3}{2n+1} - 2 \right| < \epsilon.$$

Instead a new and initially unintuitive method is adopted. First, show that the definition applies to the convergence of simple sequences, for instance when $s_n = c$ (a constant), then $s_n \to c$, and when $s_n = 1/n$, then $s_n \to 0$. (The proof of these eminently self-evident results often raises an eyebrow of suspicion from students who fail to understand them.) Next a general theorem is proved, to say that if certain sequences tend to certain limits, then the sum, difference, product and quotients of the same tend to the obvious corresponding limits. (This theorem is "obvious," but its proof, in terms of unencapsulated $\epsilon - N$ processes, imposes enormous cognitive strain for very little apparent gain.) The theorem is then quoted to show that

$$\frac{4n+3}{2n+1} = \frac{4+3/n}{2+1/n} \to \frac{4+0}{2+0} = 2.$$

Thus it is that the new types of procedure cause great difficulties. The idea that $\lim_{n \to \infty} s_n$ is both a number, and a process, and that to calculate the number requires transition to process, thence manipulation using known facts and a general theorem to get the required result, is a type of proceptual thinking that once again shuts out the procedural thinker who cannot encapsulate the limit definition as a concept through the dual meaning of notation.

It is in fact far worse, as any teacher at college level will know. The full-blown formalism of definitions of limits and axioms for the real numbers (including completeness) requires the learner to construct the properties of these defined objects by logical steps. This construction must be performed in a mind that already contains images of these properties, linked not to the definition, but to the students's previous experiences. The subtlety of performing such constructions when many of the results, as exemplified by the student's mental images, seem already to be known, causes great confusion to the majority. It is actually made worse by teaching that acknowledges this difficulty and tries to be "more informal" with the

mathematics, for the division between formal necessity and informal knowledge then becomes even more blurred.

In the case of the limit concept, students have an intuitive dynamic imagery which is in some ways in conflict with the formal definition (Tall and Vinner [18]; Cornu [2]). (For instance, they may believe that the sequence "gets closer" to the limit but "never reaches it.") Compounding this difficulty is the immense problem of manipulating several quantifiers in the formal definition. The procedural thinker who attempts to handle the definition of a sequence in a procedural manner is faced with so much detail that, once again, failure, if not inevitable, is highly likely.

Thus there is a qualitative difference between different kinds of thinking processes. In school, proceptual thinkers develop a natural knowledge structure that, of its nature, generates new knowledge with little effort; procedural thinkers seek instead the security of being able to do the processes of the mathematics, which often remains the sole focus of their effort. At college there are additional layers of sophistication which make the division between success and failure an even greater chasm.

Traditional teaching techniques usually focus on the procedural side, with the short-term aim of being able to do the mathematics. Once the procedures are sufficiently routinized to be able to carry them out almost subconsciously, they become the possible focus of reflective thinking and encapsulation as manipulable objects. It is possible therefore for a more able procedural thinker eventually to begin to compress procedural knowledge and to be able to reflect on it to move towards the required encapsulation. But the large cognitive structure required to carry out the procedure as a process in time mitigates against this success for many students. It may be a case of "not being able to see the wood for the trees," the cognitive complexity of the process completely overwhelming the conceptual simplicity of the (as yet unencapsulated) concept.

6 Using the Computer to Develop Proceptual Thinking

How then, given the divergence between more successful proceptual thinking and procedural thinking that is likely to fail, can we begin to address the growing divide? Though one might formulate a policy of making explicit the very things that the more able do implicitly (using the symbolism flexibly as both process and concept, linking together different aspects of the concepts in flexible ways), there is an inherent difficulty. If the processes are not encapsulated, then coordinating processes occurs in time, it involves more low level detail, and it imposes greater cognitive strain on an already stressed individual. Failure seems inevitable.

However, the new technology gives a new and powerful facility. If the procedure can be automated as a computer algorithm, and virtually all of them can, then it may be possible for the computer to carry out the procedure, relieving the individual of the cognitive stress of coordinating the detail, and allowing the individual to concentrate on the relationships between the mathematical objects produced by the process. At one time the individual may concentrate on the procedure and what that entails, without thinking about higher level relationships involving the product of the procedure. At another time the individual may use a computer to carry out the procedure and concentrate on the relationship between the concepts. I term this the principle of selective construction (Tall [15]).

Whilst the traditional approach dictates that familiarity with the procedure must come before reflection on its product, using the computer there is sometimes a choice as to which may be done first. Sometimes the product of the procedure, if meaningful, may be explored before the procedure is practiced and interiorized by the individual, sometimes the procedure may be practiced before the product is studied in detail. This therefore gives new possibilities for curriculum sequencing.

For instance, in Tall and Thomas [17] we attacked the problem of giving meaning to algebraic notation by using programming in BASIC to calculate the values of algebraic expressions for given values of the variables, to see how the symbolism had a certain consistency and that different looking expressions (such as $2 * (x + y)$ and $2 * x + 2 * y$) always gave the same numerical result. Because the computer was carrying out the process of calculation, the student could concentrate on the products and see that $2 * (x + y)$ and $2 * x + 2 * y$ are equivalent expressions although as processes of computation they are different. We used a simple piece of software that accepts standard algebraic notation (with implicit multiplication and powers given by superscripts) to allow the students to see that $2(x + y)$ and $2x + 2y$ are likewise equivalent (Figure 1). In Tall [12] a similar experiment was done with the visual beginnings of the calculus in the

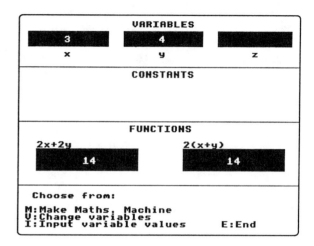

Figure 1: Equivalent outputs of different processes.

English sixth form (Senior High School, ages 16–18), and showed a great improvement in the meaning of the derivative concept. By visualizing the gradient of a graph as the gradient of a highly magnified (locally straight) small part of the graph, the student learned to look along a graph to see the changing gradient. The computer could draw an accurate representation of the numerical gradient $(f(x + h) - f(x))/h$—carrying out a process that the student could not do with such precision—and the student could look at the graph of the gradient, see how it stabilized as h becomes reasonably small, and conjecture what the gradient curve might be approximating to. At the college level, Heid [9] has shown that a combination of conceptual learning at one stage using a symbol manipulator to carry out the routines of differentiation and later practice at the procedures of differentiation produced far more flexible and versatile learning.

The new calculus curriculum in the UK designed for 16- to 19-year-olds by the School Mathematics Project uses software to visualise gradients and to guess the formulae of gradients of graphs before discussing the symbolic procedure of calculating limits. It also uses software to enable the learner to physically construct an approximate solution of a first order differential equation as the reverse process of knowing the gradient of a graph and using software to build up a curve which has the required gradient. This is done before considering any numerical or symbolic method of solution.

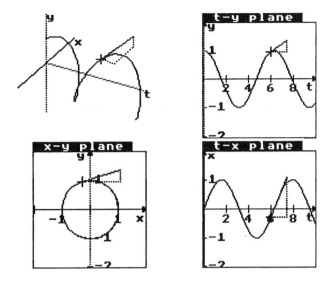

Figure 2: The numerical tangent to the curve
$x = \sin t$, $y = \cos t$ in 3D and the
projections onto coordinate planes.

The differentials dx, dy are visualized as coordinates of the tangent vector to the curve $y = f(x)$ in the x-y plane and in three dimensions the tangent to the curve $y = y(t)$, $x = x(t)$ has components (dt, dx, dy) (Figure 2). This allows a differential equation in several variables to be given a physical meaning (as specifying the direction of the tangent) and allows a formula such as

$$\frac{dy}{dx} = \frac{\frac{dy}{dt}}{\frac{dx}{dt}}$$

to be given a meaning as ratios of the components of the tangent vector. Figure 2 shows the tangent vector (calculated as a numerical approximation) drawn by the Parametric Analyser (Tall [14]). In these experiments we see visualization playing an increasing role in conceptual learning. In addition to the symbols representing both process and concept, they are also linked to other representations, such as graphical representations which expand the flexibility of conceptualization. The good mathematician evokes whichever representation is appropriate for a particular purpose, and uses that representation as long as it proves successful, switching to another representation when it proves more useful. This flexible form of thinking is often termed "versatile thinking" (Tall and Thomas [16]). Robert and Boschet [10] show consistently that the most successful students in advanced

mathematics are those with the flexibility to work in more than one representation (graphic, numeric or symbolic). Those who are limited to one representation (usually numeric or symbolic) are less likely to solve a wide range of problems.

For example, if a student consistently relies on symbolic manipulation without other representations, how will that student respond to a symbol manipulator which gives the response:

$$\int_{-1}^{+1} \frac{1}{x^2}\,dx = \left[-\frac{1}{x}\right]_{-1}^{+1} = -2.$$

7 The Achilles Heel of a Symbolic Manipulator in Education

Symbol manipulators were originally designed to enable computers to execute many algorithms in advanced mathematics. Initially it was far more a problem of getting the computer to jump through the appropriate hoops than it was of thinking of the eventual educational use of the software. Now that symbolic manipulators are increasingly flexible and stable software environments, they are becoming more appropriate for teaching and learning as well as for research.

Initially manipulators were based on the teletype interface in which a line of symbols is typed and a response is given by the terminal. At the present time most remain environments where lines of symbols are input and evaluated, though the output may now include graphs as well as symbols (which may be printed in standard mathematical notation). Thus a line of symbols, conceived either as an expression to be evaluated or a process to be carried out, is given to the computer for processing. Essentially the software accepts a procept which it processes internally, performing a construction for the user and giving a response for the user to interpret. It performs one aspect of selective construction: that of carrying out the internal procedure.

More flexible interfaces are being developed offering other forms of communication. For instance, *Mathematica* and *Maple* allow a graph of a surface (initially input symbolically) to be pulled around, using a mouse to select the viewing point rather than requiring the coordinates explicitly. But the basic mode of communication remains symbolic input, internal processing and symbolic or graphical output.

What must be apparent from the discussion in the previous sections is that the use of manipulators demands a proceptual understanding of the symbolism involved. Thus, in education, the question is whether procedural thinkers can benefit from the use of symbolic manipulators, and whether the manipulator can be used in a wider educational context to promote more flexible proceptual thinking. The Achilles heel of the symbolic manipulator in education is the need for the individual to construct a meaning for the symbolism as flexible process and product and the fact that symbolic manipulators process input internally in a manner that may not be transparent to the user. The mere surface manipulation of symbols is not enough.

Given the worsening ability of students with algebraic manipulation (certainly in Britain where there is now more emphasis on numerical problem-solving than on algebra) the need to give meaning to the symbolism becomes even more important. In the UK teachers are finding that beginning calculus students are less likely to be able to find the local maxima and minima of cubics because, although they can differentiate the expressions, they cannot factor the resulting quadratic. The latter could, of course, be performed trivially by a symbolic manipulator, but if the symbolism has little meaning, what use will this be?

If the students are able to give some meaning to the symbolism, then manipulators can be of definite assistance. For instance, in a course at college calculus level for student teachers at Warwick University, the accent was placed on visualizing the concepts first. The students knew the meaning of differentials dx and dy as components of the tangent vector. They could see, in a three-dimensional picture, the meaning of partial derivatives through taking cross-sections of a surface and looking at the gradient of each. They could do simple differentiation and integration in a meaningful way. But when it came to finding the maxima and minima on a surface $z = f(x, y)$, the sheer drudgery of working out second derivatives in certain examples defeated them. Here the use of a symbol manipulator (*Derive* and *Mathematica* were both used) proved to be greatly illuminating. The initial calculations are shown in Figure 3. It proved a simple matter to get the software to calculate the second partial derivatives and check the required conditions. The use of notebooks in *Mathematica*, which give electronic text whose symbols may be selected, evaluated, modified and investigated, promises greater flexibility for the active learner, although this

```
In[1]:= z=(a x + b y + c)^2/(x^2+y^2+1)

              (c + a x + b y)
Out[1]=     ------------------
                 2     2
              1 + x  + y

In[2]:= p=D[z,x]

           2 a (c + a x + b y)     2 x (c + a x + b y)
Out[2]=    ------------------- - ----------------------
               2     2                2     2  2
            1 + x  + y             (1 + x  + y )

In[3]:= q=D[z,y]

           2 b (c + a x + b y)     2 y (c + a x + b y)
Out[3]=    ------------------- - ----------------------
               2     2                2     2
            1 + x  + y             (1 + x  + y )

In[4]:= r=D[z,x,x]

                 2
               2 a            8 a x (c + a x + b y)
Out[4]=    ------------- - ---------------------- +
             2     2           2     2  2
          1 + x  + y        (1 + x  + y )

           2             2                       2
         8 x  (c + a x + b y)     2 (c + a x + b y)
        --------------------- - ---------------------
             2     2  3             2     2  2
          (1 + x  + y )          (1 + x  + y )
```

Figure 3: The initial calculations of a max/min problem with *Mathematica*.

still must be done within the syntax and facilities allowed by the software, with the internal procedures hidden from sight. But it should be remembered that the relationships generated by such manipulators work only in certain ways, for instance, from symbolic input to graphical output. Other directions, for instance, using graphical concepts to produce related symbolic notions, still need to be done by the human mind.

Using the principle of selective construction, one may hypothesize that the manipulator must be used as part of a wider educational context in which active learning is encouraged to develop flexible use of symbols as procepts and to link these symbols to other representations. This needs to be performed within a wider educational framework that encourages the student to develop a flexible understanding of the symbolism.

8   The Need for Versatile Learning

Symbols alone cannot provide a total environment for mathematical thinking. They must represent something, and are more powerful if they do so in a flexible proceptual way. The power is further enhanced if there are alternative representations available which increase the flexibility of thinking.

Tall [13] analyses the content of the calculus syllabus from a cognitive viewpoint, concentrating on the processes of differential and integral calculus. To my surprise, I found that the basic cognitive concepts were not differentiation and integration. Instead, I found that I needed to start with the more fundamental notion of change and see differentiation and integration as the symbolic parts of rate of change and cumulative growth.

The notion of change is represented by the function concept, which may now be seen as a flexible procept. It can be carried out as a process of assignment, it can be reversed as an inverse function, or as a solving of equations. It has several different representations, of which I concentrated on three: the symbolic, the numeric and the graphic. The symbolic and the numeric are both proceptual. They are procedures for calculating values of the function, which can also be conceived as objects (as expressions or as named computer procedures). The graphic representation (a function as a graph) occurs in a symbolic manipulator only as the output of a numerical procedure (which might be specified symbolically). As we saw earlier in our analysis of the difference between geometry and algebra, the visual concept tends to be seen as an object—a curve in space—rather than a process (take the value of x on the x-axis, move up to the curve $y = f(x)$ and across to the y-axis to find the corresponding function output). This is known to be a weakness of the graphical representation. But the graph also gives a large amount of qualitative information that enables the user to conceptualize global concepts that are often hard to imagine purely from the symbols or numbers. It therefore occupies a worthy place alongside symbolism and numeric procedures as representations of the fundamental notion of change.

Differentiation occurs as the symbolic part of rate of change, and integration as the symbolic part of cumulative growth. Each of these notions occurs as a process which can be done, and undone. The doing in each case is, of course, a procept, and the undoing is the reversal of the process part, which has a complementary proceptual structure. An interesting facet of this conceptual analysis is that the undoing of differentiation is not integration. It is the solving of a differential equation. In visual terms the qualitative idea of the derivative is the gradient of a graph, which may be seen by looking at the graph under high magnification so that it looks "locally straight." It is possible at a primitive level to see the gradient of (the graph of) a function simply by casting one's eye along and estimating the changing slope. Once this is established and one can see a number of standard formulae (such as the derivative of x^2, x^3, x^n, $\sin x$, $\cos x$, $\ln x$, e^x, etc.), the qualitative picture becomes an encumbrance and one develops more powerful ways of calculating the gradient through rules of symbolic manipulation. Thus it is that the process of enaction of the gradient to be later replaced by the use of symbolic manipulation resembles earlier encounters with procepts. First it is necessary to give the concept a meaningful representation. Then aspects of the representation itself start to take on a life of their own (in this case the symbolic process of calculating a derivative) and the learner need only build on the meaningfully encapsulated procept (symbolically calculating the derivative), rather than stepping all the way back to first principles. The same is true of young children counting. At first they need physical objects to operate on, but when the symbols have meaning, it is only necessary to depend on the meaningfulness of the symbols.

Solving a differential equation is the reverse of the process of finding the gradient: in primitive terms it is a matter of knowing the gradient everywhere and trying to build up the graph. Many pieces of software are available to draw direction diagrams and to draw numerical solutions automatically. The Solution Sketcher (Tall [14]) allows the user to experience the physical act of building up a solution with the computer using the (symbolic) first order differential equation to calculate and draw a small line segment of the appropriate gradient through a selected point in the plane (Figure 4). Although this picture seems fairly innocuous, it is a potent enactive environment which enables the user to point anywhere in the plane and deposit a small line segment whose gradient is given by the differential equation. By putting such segments end to end and leaving a trace on the screen, the student can build up a solution curve. I see this enactive process of building up a solution curve as a fundamental physical action which gives a primitive meaning to the differential equation. Yet it forms the cognitive foundation of more formal concepts, such as the uniqueness of a solution through a given point (provided that the gradient is properly defined). Once the student has internalized the meaning of a solution of a differential equation in this way, it is soon apparent that the solution is found more easily either by

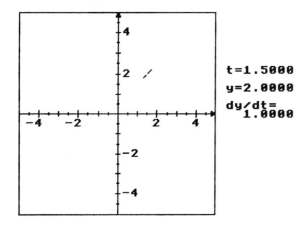

Figure 4: The Solution Sketcher ready to build up the solution of a differential equation.

using various numerical procedures, or through attempting to reverse the methods of symbolic manipulation. Integration, as the area under a graph, or Riemann integration, may be classified as a cognitively different structure: cumulative growth. It has its own symbolic, numeric and graphic interpretations. Symbolic integration as a theory proves to be of great interest because of the fundamental theorem of calculus, which shows that it can be carried out by anti-differentiation. In practice the graphic and numeric interpretations of undoing cumulative growth are little studied because of the overwhelming power of the fundamental theorem.

This analysis of the structure of the calculus is given in Figure 5. In this picture, the traditional part of calculus, the symbolic part, is but part of a much larger picture. This part is most valuable for its ability to carry out the procedures so successfully, where a picture might only give a qualitative idea and a numerical procedure only an approximation. But the symbolic undoing of differentiation, differential equations, involves many problems for which symbolic methods do not give a solution (in terms of elementary functions). Thus the wider picture, in terms of the practicality of solving problems and the visualization of the qualitative concepts, takes on an important overall role.

It should be very strongly emphasized that the existence of this structure does not mean that all the parts of the diagram need to be given equal weight, but that they should be used for their appropriate purpose. For instance, pictures should be used for conceptual insight, whilst numerical calculations or symbolic manipulation are used for productive calculation. The good mathematician selects whichever representation is appropriate for a given stage of a given problem, moving flexibly between representations where this becomes expedient. It is the versatility to move between representations and choose the most appropriate that gives the good mathematician great power. Such an approach does not overburden short-term memory by working on several different representations simultaneously. The desire to coordinate several different representations and to see processes carried out simultaneously in all of them can easily overstretch the working mental capacity. If it strains the good mathematician, it is even more likely to overburden the average student. What is more important is to allow students to perform more in the mode of a good mathematician by allowing them to selectively construct part of the conceptual structure that is the current focus of attention whilst the computer carries out other parts of the constructive process.

9   Summary

Symbolism is used flexibly by the good mathematician. Symbols allow mathematical thinking to be compressible, so that the same symbol can represent a process, or even a wide complex of related ideas, yet be conceived also as a single manipulable mental object. This flexibility is stock-in-trade for the mathematician. But it is not for the average student, who seeks a shorter-term goal: to be able to do mathematics by carrying out the necessary processes. It is this relationship between procedures to do mathematics and the encapsulation of these concepts as single objects represented by manipulable symbols that is at the heart of mathematical success and its absence is a root cause of failure.

We therefore see that the use of symbols in a wider sense, dually representing processes or concepts, linked with other representations including visualizations, gives a flexible view of mathematics that makes the subject easier for the more able. The less successful tend to cling more to a single representation, often a procedurally driven symbolic approach, which is

Representations:		
Graphic	**Numeric**	**Symbolic**
Qualitative Visualizing Conceptualizing	**Quantitative** Estimating Approximating	**Manipulative** Formalizing Limiting

Concepts:

Change

	Graphic	Numeric	Symbolic
doing:	graphs	numeric values	algebraic symbolism
undoing:	graphical solutions of equations – intersection of graphs	numerical solutions of equations – sequences of numerical approximations	inverse functions (solving equations) symbolic solutions

Rate of change

	Graphic	Numeric	Symbolic
doing:	local straightness	numerical gradient of graph	derivative
undoing:	build graph knowing its gradient	numerical solutions of differential equations	solutions of differential equations –antiderivative

Cumulative Growth:

	Graphic	Numeric	Symbolic
doing:	area under under graph	numerical area	integral
undoing:	know area –find curve	know area – find numerical function	FUNDAMENTAL THEOREM

Figure 5: The conceptual Structure of the Calculus

inherently less flexible and imposes greater cognitive strain on the user. The short-term gain of showing a student the procedure to be able to do a piece of mathematics may, for these students, lead to a cul-de-sac in which security in the procedure prevents the flexible use of symbolism as both process (to obtain a result) and object (to be able to manipulate as part of higher level thinking). Now that computer environments are available to carry out algorithmic processes in a predictable manner, it may be possible to encourage a wider range of students to gain flexible insights into the higher level concepts, integrating them in a more proceptual manner, linking to other representations.

Symbolic manipulators, taking a proceptual input and internally carrying out procedures that are usually invisible to the user, may be used to complement the skills of the student, but this requires some insight into the meaning of the symbolism. Therefore the manipulators are better used as part of a richer environment which helps the students develop supportive linkages between concepts. They can provide an environment for manipulation of such symbolism, by carrying out the process and enabling the user to focus on the concept. This principle of selective construction offers a method of reducing cognitive strain and increasing the student's chances of developing more flexible thinking processes.

References

1. Beth, E. W. and Piaget, J. (1966). *Mathematical Epistemology and Psychology* (W. Mays, trans.), Dordrecht: Reidel.

2. Cornu, B. (1981). "Apprentissage de la notion de limite: modèles spontanés et modèles propres," *Actes du Cinquième Colloque du Groupe Internationale PME*, Grenoble, 322–326.

3. Dubinsky, E. (1991). "Reflective Abstraction," in D. O. Tall (ed.) *Advanced Mathematical Thinking*, Dordrecht: Reidel.

4. Freudenthal, H. (1983). *The Didactical Phenomenology of Mathematics Structures*, Reidel, Dordrecht.

5. Gray, E. M. (1991). "An Analysis of Diverging Approaches to Simple Arithmetic: Preference and its Consequences," *Educational Studies in Mathematics*, (in press).

6. Gray, E. M. and Tall, D. O. (1991). "Duality, Ambiguity and Flexibility in Successful Mathematical Thinking," *Proceedings of PME XIII*, Assisi.

7. Greeno, J. (1983). "Conceptual Entities," in D. Genter and A. L. Stevens (eds.), *Mental Models*, 227–252.

8. Harel, G. and Kaput, J. J. (1991). "The role of conceptual entities and their symbols in building advanced mathematical concepts," in D. O. Tall (ed.), *Advanced Mathematical Thinking*, Dordrecht: Reidel.

9. Heid, K. (1984). *Resequencing skills and concepts in applied calculus through the use of the computer as a tool*, Ph.D. thesis, Pennsylvania State University.

10. Robert, A. and Boschet, F. (1984). "L'acquistion des débuts de l'analyse sur R dans un section ordinaire de DEUG première année," *Cahier de didactique des mathématiques* 7, IREM, Paris VII.

11. Sfard, A. (1991). "On the dual nature of mathematical conceptions: reflections on processes and objects as different sides of the same coin," *Educational Studies in Mathematics*, 22, 1, 1–36.

12. Tall, D. O. (1986). *Building and Testing a Cognitive Approach to the Calculus using Computer Graphics*, Ph.D. Thesis, Mathematics Education Research Centre, University of Warwick, UK.

13. Tall, D. O. (1990). "A Versatile Approach to Calculus and Numerical Methods," *Teaching Mathematics and its Applications*, 9, 3, 124–131.

14. Tall, D. O. (1991). *Real Functions and Graphs*, software for BBC compatible computers, Cambridge University Press.

15. Tall, D. O. (In press). "Interrelationships between mind and computer: processes, images, symbols," *Advanced Technologies in the Teaching of Mathematics and Science* (David L. Ferguson, ed.), Springer-Verlag.

16. Tall, D. O. and Thomas, M. O. J. (1989). "Versatile Learning and the Computer," *Focus on Learning Problems in Mathematics*, 11, 2, 117–125.

17. Tall, D. O. and Thomas, M. O. J. (1991). "Encouraging Versatile Thinking in Algebra using the Computer," *Educational Studies in Mathematics*, 22, 2, 125–147.

18. Tall, D. O. and Vinner, S. (1981). "Concept image and concept definition in mathematics with particular reference to limits and continuity," *Educational Studies in Mathematics*, 12, 2, 151–169.

Part II:

Symbolic Computation in Calculus

Using a Symbolic Computation System in a Laboratory Calculus Course

L. Carl Leinbach
Gettysburg College

1 Introduction

There are several good articles discussing symbolic computation systems that are available for classroom use. An excellent example is Simon's article (see [3]). These articles tend to discuss the performance characteristics of the systems and compare their performance on a set of test problems. The information contained in these articles is very useful in helping a department to choose a system, but this choice should only be made after deciding the central issue of how the department plans to use the system in teaching its courses. In this article I will consider the calculus course and propose a model for the use of a symbolic computation system in this course. My premise is that the use of a symbolic computation system should change our perspective on teaching undergraduate mathematics.

At the present time much of the use of these systems is as a way to avoid tedious and annoying aspects of our daily chores or as a way to include more complex and seemingly realistic applications in our courses. These uses of a system are not bad and may, in fact, be exciting to many of our students. There is, however, an even more exciting possibility. This possibility is to use the system as a way of inviting our students to act as practicing mathematicians. To lead and encourage them to make conjectures, test their conjectures, and, ultimately, to become involved in their verification. This is a possibility that is far removed from the student's role in the traditional method for presenting calculus courses.

The use of a symbolic computation system makes it possible to experiment and observe mathematical phenomena. In order to achieve maximum effect at this level the experimentation needs to be directed and the student's powers of observation honed. I propose that we teach our calculus courses as true laboratory courses in the style of the introductory courses taught by our colleagues in the natural sciences.

I use *Derive* in the laboratory, but that is not the important point. I use it because I believe that it is the best system for my environment. However, the course can be taught using any of the current systems that include competent symbolic manipulation, accurate numerical approximation, and good two-dimensional graphics. The calculus laboratory is not system dependent or machine dependent.

2 A Calculus Course with a Laboratory

According to *The American Heritage Dictionary* [1], the epistemology of the sciences consists of five steps: observation; identification; experimental investigation; analysis; and explanation of natural phenomena. It is this process that I advocate for the investigation of mathematical phenomena. The symbolic computation system allows the properly directed student to perform the first three steps of the process within the context of a scheduled laboratory. The last two steps are done in the classroom and as part of homework assignments. The laboratory experience gives the student the background and intuition to operate as the instructor's mathematical colleague in the sense that the student has been engaged by the subject and is proposing avenues of exploration and searching for explanations of observed phenomena.

Adding the requirement that students work with

lab partners places them in a comfortable situation for discussing mathematics with their peers. The phenomena they observe are not immediately obvious. A typical lab conversation is: "What are we supposed to be seeing here?" "I'm not sure. Let's look at the lab write-up again." "What if we try this." Not all conjectures are correct or even sensible from our perspective. However, students should feel comfortable in making conjectures and attempting to justify them. If a conjecture is incorrect the student must be encouraged to explore counterexamples and reformulate the conjecture. The instructor also has an obligation to understand why a student was led to a particular conjecture and help that student develop the maturity to make reasonable observations.

The explanation and justification of a conjecture is done in the laboratory report. This is an important part of the laboratory. There are two reasons for this statement. The first is that the struggle to state a conjecture in a clear, concise way forces students to evaluate it. Is it reasonable? Is it saying anything of substance? Is this really what I saw? Am I saying what I mean? The second reason is that students learn to discuss mathematics in a natural language. It dispels the thought that only the "initiated" can discuss mathematical ideas. Even though the proof of a conjecture may require sophistication and mathematical insight, it is possible to discuss its significance with an audience that extends beyond the traditional mathematical community.

Given the claims for these general benefits, how exactly does one go about conducting a lab? The following are examples of labs that can be done using a symbolic computation system. All of these labs are used in our calculus classes at Gettysburg.

3   The Laboratory Period

The calculus class meets three periods per week for lecture and discussion and a fourth period for the laboratory. The laboratory period is seventy-five minutes long and is conducted in the microcomputer lab. Students work in pairs and each pair is seated at an MS/DOS machine running *Derive*. There is a laboratory procedure for each lab that is given to the students prior to the lab and it is expected that the students have read this procedure when they come to lab. A brief introduction to the lab is given using a projection panel and computer similar to the student machines. The introduction contains a review of the

key strokes necessary to set up a screen that can display both text and graphics and a survey of the lab for the day. After the introduction the students follow the lab procedure and fill out the lab work sheet that they will turn in at the end of the lab period. The instructor circulates about the lab and answers individual questions.

Some of the labs deal with applications of the calculus to reinforce the fact that computation can assist the student in the problem solving process. This type of lab is inserted at the time when applied max/min problems, related rates, and applications of the definite integral are being studied. In this case the lab material follows the classroom presentation and, in most cases, the student duplicates the process followed by the instructor in class. The more common format for a lab is one that anticipates material to be presented in class.

We use three types of labs to anticipate results and encourage students to make conjectures. These are labs that are descriptive of the behavior of functions; labs that urge the students to recognize and search for patterns produced by symbolic computation; and labs that anticipate important conceptual results.

4   Describing the Graphical Behavior of Functions

It is our experience that first-year college students have very little graphical intuition and do not readily relate an algebraic expression to a graph. A prime reason for restructuring the calculus course as a laboratory course is to help develop this intuition. Starting with the first lab, students are trained to look at the graph of a function and interpret what they see.

While students are learning to use the *Derive* facilities for function definition, entry of expressions, and plotting, they examine the graphs they generate. Given the graph of $f(x)$, they also look at the graph of $f(x + a)$ for specific values of a. This is done for at least five different functions and four values of a. They are asked to make a conclusion about the effect of this transformation on the graph of the function. What aspects of the graph remain constant? What aspects change? The students also look at the graph of $f(x)+a$. Finally, they plot the following set of data points (as in Figure 1) and find a function that passes through the points. As part of the earlier exercises the students had examined the graphs of the sine and cosine functions.

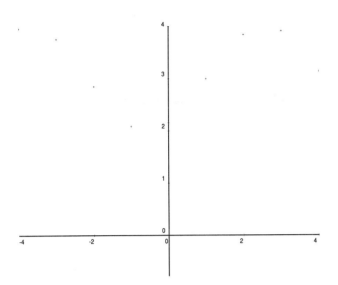

Figure 1: Points on a graph.

x	-4	-3	-2	-1	0	1	2	3	4
y	3.96	3.76	2.85	2.09	2.16	3.00	3.84	3.91	3.14

Almost all of the students recognized the shape of the sine function (some tried a cosine function). After some discussion, the translation of 3 units up was tried and then the translation of 1 unit to the right was found, yielding $\sin(x-1)+3$.

Another exercise of this type that plays an important role in the course is to observe the relationship between a function and its derived function. This exercise is very familiar to those who know the folklore of symbolic computation systems. In the lab approach the students do the exercise prior to any formal discussion about the significance of the sign of the derivative. The student explores the graphs of several function-derivative pairs such as the graph of Figure 2 for $f(x) = x^3 - x + 1$.

Sign of $\frac{d}{dx}f(x)$	Behavior of $f(x)$
Zero	*One local maximum, one local minimum*
Positive	*Increasing in both cases*
Negative	*Decreasing*

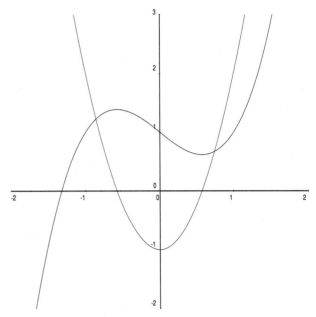

Figure 2: $f(x) = x^3 - x + 1$ and $f'(x)$.

For each of the function-derivative pairs the student completes a chart such as the one shown above. The italicized entries are the desired student responses. In the lab report the student is expected to make a conjecture about the sign of the derivative and the behavior of the function. The student is also asked to give a general reason why this behavior should be expected. The success in complying with this last request is varied. It generally works best in classes where the idea of the derivative as the instantaneous velocity has been discussed.

5 Discovering Some Symbolic Computation Algorithms

This lab period begins with the student recording the results of applying the differentiation operator to some polynomials, $\sin(x)$, $\cos(x)$, $\tan(x)$, $\arctan(x)$, and e^x. The student records these results on a lab sheet and is directed to apply the operator to several sums and differences of these functions. One example may be: $x^3 - 5x + \sin(x)$. The student records the results and is asked to identify a general pattern for $\frac{d}{dx}[f(x) \pm g(x)]$. The same procedure is repeated for examples of the form $f(x)g(x)$ and then for examples of the form $f(x)/g(x)$. Assuming that the instructor has carefully chosen the examples, the students can recognize the product rule pattern. Very few students recognize the pattern for the quotient rule. The reason is that *Derive* does not display the result in its canonical form. It simplifies the result and displays the sum of two fractions. In short, it treats the quotient as the product $f(x)(1/g(x))$. Within the lab report most of the students can prove the rule for the derivative of the sum and difference of functions. Fewer students are able to prove the product rule for derivatives, and virtually none of the students are able to prove the quotient rule. A few enterprising students did check their text for the results. The next form of the lab will ask the student to investigate the derivative of $1/g(x)$ before attempting the quotient rule.

Another lab of this type is one that is done shortly after studying integration by parts. The student applies the integration operator to the following functions: $x^n e^x$; $x^n \ln(x)$; $[\ln(x)]^n$; and $x^n \sin(x)$. After observing the results for $n = 1, 2, 3$, and 4, the student is asked to find a general pattern for general n. Some of these patterns will be recursive. Since the general result can be checked using the symbolic computation system, students are asked to review mathematical induction and to justify their observed patterns.

6   A Laboratory that Anticipates a Major Result of the Calculus

So often the first part of the Fundamental Theorem of Calculus is ignored by students in favor of the more immediately appealing computational statement of the corollary of the theorem:

$$\int_a^b f(x)\,dx = F(b) - F(a)$$

where $F'(x) = f(x)$.

Granted, they are learning to use a very powerful computational tool, but the inverse relationship that exists between the accumulation of functional values and the rate of change is missed by this emphasis. The lab that is scheduled immediately prior to the class presentation of the Fundamental Theorem of Calculus deals with the definition of the definite integral.

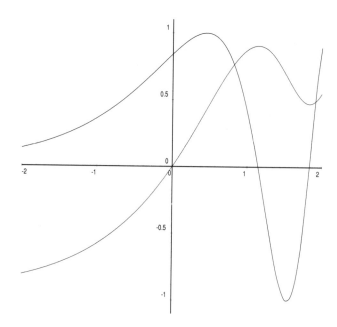

Figure 3: $\sin(e^x)$ and $S(0, x, 10)$.

Students begin by entering the following definition:

$$S(a, b, n) := \frac{b - a}{n} \sum_{i=1}^{n} f\left(a + \frac{(b - a)}{2n} + \frac{(i - 1)(b - a)}{n}\right).$$

This is, of course, the midpoint approximation of the definite integral using a partition of n equally spaced intervals. After defining $f(x)$, this definition and *Derive*'s approXimate operator are used to estimate the value of some definite integrals that the students had done as part of a homework assignment. This in itself is impressive, but not the goal of the lab.

The next item on the agenda is to fix n at 10 and consider $S(0, x, 10)$. A table of values for $S(0, x, 10)$ is constructed. This table emphasizes the fact that $S(0, x, 10)$ is a function of x. But, what does the graph of the function look like, and what is its relation to $f(x)$? These questions can be answered by studying the graph displayed in Figure 3 together with the table of $S(0, x, 10)$ (labeled S in the table) given below.

x	0.00	0.25	0.50	0.75	1.00	1.25	1.50	1.75	2.00
S	.000	.226	.473	.710	.876	.890	.714	.489	.546

During the course of the lab period, the students observe several graphs and the example given above is one of the last that they consider. Other examples include polynomials, the sine function, and one having a jump discontinuity. By the time of this lab the students are conditioned so that they can observe that $f(x)$ is a good approximation of the derivative of $S(0, x, 10)$. A close examination of the graph will show that $f(x)$ is "not quite" the derivative of $S(0, x, 10)$. The important point is that students can observe a relationship that is within the context of their experience in calculus. They are prepared to consider the Fundamental Theorem and have a stake in its proof.

7 Discussion

The laboratory approach to teaching the calculus course has much to recommend it. Students are invited to make observations and to use the insights that they gain to make explanations and conjectures. It can be argued that not all results in mathematics are intuitive and that the use of this approach can lead a student to an incorrect conclusion. This is always a possibility, but at this level such results are the exception and not the rule. It is also for this reason that the student should come to appreciate the fact that no result in mathematics can be accepted until a proof has been given. This is the point at which the mathematician and the laboratory scientist part company. Ulam [3] quotes Dirac as saying "Show me some pretty equations and then, if they are nice enough, go and prove that they are true." Surely, most practicing mathematicians do not subscribe to this agenda. The laboratory approach is a compromise between this extreme and the one in which the student has little intuition to use as a guide. By exploring some carefully chosen examples, it is possible to make conjectures about general behaviors.

Others may argue that one should not trust computer graphics or symbolic algorithms. This argument also has merit. For example, when graphing $f(g(x))$ for $f(x) = x^2$ and $g(x) = \sqrt{x-4}$, *Derive* graphs a straight line with no regard to the fact that the domain is restricted to $x \geq 4$. In such cases we should ask students to comment on the graphic output that is generated. Another type of graphing problem occurs when one considers the graph of a function that has roots located very close to each other. The observer may not see one of the roots. Another type of problem occurred when graphing the branches of the solution to the equation $x^2 - 2xy + y^2 - x - 4y = 3$. *Derive* quickly determined that $y = x + 2 \pm \sqrt{5x + 7}$, but when graphing the two functions, a gap appears in the vicinity of $(-1.5, .6)$. The reason for this peculiar behavior of the graphic display is not easily explained to first-year students.

Other problems may arise with the symbolic computation algorithms themselves. Earlier it was noted that the quotient rule does not give results in the same form as a calculus textbook. The same problem arises when doing symbolic integration. The substitution covered in class may give a result that appears to be different from the one obtained by symbolic computation. In other cases the system may fail to do an integral or solve an equation that seems to be fairly straightforward. These problems are not substantial and are explainable, but they are confusing to the first-year college student. For this reason, all lab examples should be carefully pretested. Failure to do this will result in labs where more time is spent explaining anomalies than in exploring ideas.

8 Evaluation

Teaching calculus as a laboratory course is indeed more work than teaching it in the traditional way. Additional preparation time must be given to the labs. Reading and grading lab reports is a formidable task. Not all faculty teaching the calculus see the necessity for doing this extra work. They devote their time to their students in other ways. In a larger university setting, calculus teachers may be graduate assistants. Is it reasonable to expect these younger, underpaid, and inexperienced teachers to do this task? The reasonable answer seems to be no. An alternative is to have a faculty member in charge of the lab sections coordinating the labs with the course syllabus, circulating the lab manuals to the calculus teachers, and holding occasional meetings with the staff. This arrangement allows the students to benefit from the labs and the instructors to operate in a way that they feel is most effective for them. There is one constraint placed on the classroom teachers. They must be aware of what is going on in the lab and incorporate their students' insights into their presentations. But, this is a definition of effective teaching.

Students work in pairs and turn in a joint laboratory report. This ensures that the conversation about the subject of the lab continues outside of the lab.

Admittedly, there is a great deal of word processing taking place on the evening before the lab report is due. However, hastily written and poorly developed reports are evaluated as such and returned to the student. As the term continues, I receive more visits from lab partners wanting to know "what I was looking for," and we have the opportunity to continue the discussion of the lab. It is a great joy to hear the "Aha!" as we examine their data and discuss the possible relations that are contained in that data. Not all "Ahas" result in profound insights or even great reports, but the quality of many of the reports improves as the term progresses.

There is another aspect to teaching calculus as a laboratory course that is not unique to this approach, but is facilitated by it. It allows me, as a teacher, to know what it is that students don't know in a more congenial and relaxed atmosphere than a testing situation. What I am finding is that facts I assumed were part of the students common knowledge are not known by many of the students. This extends to the vocabulary of mathematics. I was surprised during an early lab to find that many students did not know what parts of the graph of a function corresponded to $f(x) > 0$, $f(x) = 0$, and $f(x) < 0$. Many of the students thought that $f(x) = 0$ meant that the graph of the function was crossing the y-axis! Surely, this topic is part of the secondary school curriculum, but for many of the students it does not make a lasting impression. During other lab periods, it became apparent that students have very little experience using the graph of a function to describe its behavior. They also had difficulty recognizing relationships between a function and functions that are associated with it, such as its derivative or one of its antiderivatives. The relationships are there, but the students have little experience in finding them. The symbolic computation system provides a means of generating several examples and the lab period provides the opportunity to explore them with purpose.

Grading the laboratory reports provides an opportunity to flag sloppy statements such as: "We have a system of four equations in three unknowns and such systems have no solutions," and "The graph of a rational function will have a vertical asymptote whenever the denominator is zero." In one case, I found that all of the students in a particular section believed that a twice differentiable function had a saddle point whenever $f'(x) = 0$ and $f''(x) = 0$. Alerted to the situation, I was able to discuss it with the class presenting examples that showed the conjecture to be false.

9 Conclusion

Does teaching calculus as a laboratory course with a symbolic computation system improve student grades and guarantee that they will preform better on comparative tests than students in a traditional course? I don't know, nor do I consider the question important. The fact is that the calculus laboratory provides students with better opportunities for learning the calculus. They have the opportunity to investigate and explore concepts. They also have the opportunity to apply these concepts to situations that are new and different from those found in most texts. When they are writing the lab report they are involved in communicating mathematics to their reader and are discussing what they are communicating with their lab partners. When I grade the lab reports I give the students feedback on how effectivly they communicate their ideas as well as commenting on the correctness of their observations. In particular, I am involved with the students and their learning of calculus in a new and exciting way. It's more work to teach calculus this way, but I believe that it is worth it!

References

1. Davis, P. (ed), (1973). *The American Heritage Dictionary* (Paperback Edition), Dell Publishing Company, New York.

2. Rich, A., Rich, J., and Stoutmeyer, D. (1990). *User Manual Derive, A Mathematical Assistant*, The Soft Warehouse, Honolulu.

3. Simon, B. (1990). "Four Computer Mathematical Environments," *Notices of the AMS*, 37, 7, 861–868.

4. Ulam, S. (1986). *Science, Computers, and People*, Birkhauser, Boston.

Some Uses of Symbolic Computation in Calculus Instruction: Experiences and Reflections

Joan R. Hundhausen
Colorado School of Mines

1 Introduction

The recent calculus reform movement ([1], [7]) has spurred many individuals and institutions to experiment with the teaching of elementary college mathematics. While striving to build a stronger conceptual base for the mathematics and/or making it more applicable, many of the experiments have involved integrating a technological component. Although the goals, standards, subject matter, and emphases of the introductory calculus course have been called into question before, even on a national scale [8], the present discussions include a new dimension: the use of emerging technologies such as computer algebra systems. The symbolic, numerical, and graphical capabilities of these systems provide opportunities for discovery learning, exploration of mathematical concepts, and treatment of real world problems in ways that have not heretofore been feasible.

In the next section I describe some uses of the HP-28S supercalculator in a freshman-level course that fits into the broader context of calculus reform. In this unique course[1] (taught with Physics Prof. F. R. Yeatts), Calculus I and II and Physics I (Kinematics) are integrated. We are particularly concerned about enabling students to build a sound mathematical base for further studies in science and engineering. Combining calculus and physics provides opportunities (often following along historical lines) for guiding the development and reinforcement of mathematical concepts and their applications. The HP-28S facilitates treatment of the graphical, numerical, and analytic aspects of those concepts.

[1]Partially supported by NSF Curriculum Development Grant #USE–8813784.

The HP-28S has been an important part of the course in regular classroom work, as well as in special workshops, and has enabled us to realize some of the clear advantages of computational aids in the teaching of calculus. We have also become aware of the potential problems that may arise if students depend too heavily on technology. Some of these are related to considerations too often neglected by mathematicians—the ultimate use of calculus as a tool in science and engineering. Thus the second part of this paper will be devoted to some reflections on the use of symbolic computation systems in calculus instruction in relation to some of the broader goals of that instruction.

2 A Successful Learning Experience in Calculus

To set the stage for the ensuing examples and discussion, it may be well to examine some learning goals. While designing a course that emphasizes conceptual understanding rather than rote learning and manipulation of formulas, we have attempted to facilitate students' ability to transfer their knowledge of calculus concepts and techniques to other contexts. From our viewpoint, a successful learning experience in calculus may be described by the statement:

> Students should have the ability to render a simple physical (or geometric or economic, etc.) situation in mathematical terms, and be able to recognize and apply appropriate mathematical tools to analyze the behavior of the system.

Implicit here, of course, are the elements of mathematical modeling, problem-solving, and analytical thinking. Success in this sense also requires that the student has understood mathematical concepts deeply enough to be able to recognize their relevance or applicability in a context removed from the calculus classroom. Similar goals were expressed eloquently by R. Weinstock in his landmark paper, [8]:

> The teaching of calculus ought to have as one of its major aims the total psychological grasp by the student of the processes, techniques, and ideas that will enable him to do something with what he has learned...

The availability of symbolic computer systems that can perform many of the operations found in calculus courses raises vital questions about how these tools will affect teaching and learning. In what ways can the use of a symbolic computation system in the teaching of calculus enhance or hinder progress toward the goals mentioned above? How should teaching materials, methods, emphases, and testing change to incorporate the use of this technology?

Although there remains a great need for research into these questions, some preliminary studies ([5], [9]) suggest that a well planned integration of a symbolic computation system into a calculus course has a positive impact. I will describe some ways in which the capabilities of such a system (in particular, the HP-28S calculator) have been used in our course to enrich problem-solving experiences and broaden student understanding. This will be followed by consideration of some of the issues that invite further study.

2.1   Integrating a Symbolic Computer System into the Course

2.1.1   A Limit Problem and Some Exploration

Consider the following problem (taken from a standard calculus text [4]), which is based upon a simple geometric construction:

> Let R be a rectangle formed by joining the midpoints of the sides of the quadrilateral Q having vertices at the points $(\pm x, 0)$ and $(0, \pm 1)$. Find the limit of the ratio (Perimeter of R)/(Perimeter of Q) as x approaches zero.

The student must model this geometric configuration (no figure is provided with the statement of the

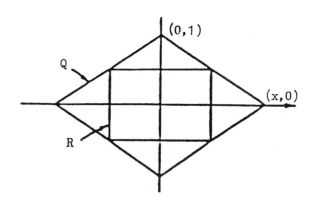

Figure 1: Rectangle R inscribed in quadrilateral Q.

problem) and derive a formula for the function $f(x)$ which represents the ratio in question (see Figure 1).

Once the formula $f(x) = (2x + 2)/4\sqrt{x^2 + 1}$ is obtained, finding the limit as x approaches 0 is straightforward and should not require a computational aid. Other aspects of this problem, however, are rather interesting, and this is where further exploration and understanding is facilitated by the calculator (or computer).

First, it is instructive for students to visualize the approach to the limit by repeated numerical evaluation of the ratio as x approaches zero. This provides concrete evidence for a rather abstract process.

Second, a graphical display of the function on the calculator will show that for very large or very small values of x, the ratio approaches $1/2$ (as is readily seen by analyzing $f(x)$). But more significantly, the display indicates that there is an intermediate value of x for which the ratio actually peaks. This phenomenon may well have been missed without the aid of the graphing facility (or perhaps the facility to perform many numerical calculations rapidly). A potentially dull and routine problem becomes more interesting; students are motivated to evoke the power of the calculus to find the precise dimensions of the configuration exhibiting the largest ratio of perimeters. This critical value of x as well as the corresponding value of the ratio can be found approximately by "digitizing" the location on the graph that appears on the calculator screen. (Scaling of the graph may be necessary for this rather broad maximum; see Figure 2.)

The symbolic and computational power of the calculator can then be used very efficiently to find the

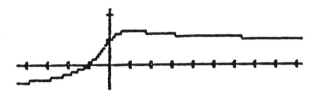

Figure 2: Magnified calculator graph of the function

precise location of the critical point. (Here it is assumed that the student has the "total psychological grasp of the processes and techniques" involved; further discussion will be offered in the second part of this paper.)

Although the hand calculations that would be necessary to complete this problem are not difficult, one can easily see how the power of the symbolic computation system can be brought to bear, with great advantage, on more complicated problems where hand calculation could be much more tedious. Finally, the easily accessible information about the global as well as the local behavior of the function is also valuable. This example suggests how the availability of a powerful system may be an aid in encouraging a spirit of exploration and discovery in the classroom and in homework; the possibilities for enrichment are many!

2.1.2 Introduction to Numerical Solution of Differential Equations

The next illustration suggests how the notions of approximation, convergence, and error analysis can be introduced with the aid of a symbolic computation system. Euler's method for numerically solving differential equations ([2], [3]) has been found to be a particularly versatile tool for developing algorithmic thinking, while reinforcing and unifying important mathematical concepts. Once students have mastered the notion of the tangent line to the graph of a known function $f(t)$, the idea can be "reversed"; the slope $f'(t)$ at a point on the graph of a function can be used to predict the approximate change in the function in response to a small change h in the independent variable.

$$f(t+h) \approx f(t) + f'(t)h,$$

or more generally,

$$y(t+h) \approx y(t) + g(t,y)h,$$

where $dy/dt = g(t,y)$.

Repetition of this prediction process, using known information only about the rate of change of a function at a point and a starting value for the dependent variable, enables the calculation and plotting of successive predicted values of the function.

The numerical and graphical capabilities of the calculator or computer combine to provide a display of approximate solution curves for various values of the step size h. This leads the student to an intuitive grasp of what is meant by solving a differential equation and convergence at a point or over an interval. In a course that integrates calculus and physics, interest in solving differential equations evolves naturally; a variety of acceleration, velocity, and distance relationships can be explored. For certain simple cases, say $f'(t) = 2t$, the student can find the true solution curve analytically, compare its graph with those of the approximating solutions, and become aware of some introductory aspects of error analysis. A specific application which has been explored early in our course is that of the trajectory of a particle with or without air resistance—and somewhat later, compound interest and Kepler orbits.

While reinforcing the concept of derivative and providing an application of the tangent line, the Euler method provides a foundation for more sophisticated approximation techniques in advanced courses. This topic is revisited throughout the course as maturity develops. For example, one-dimensional time-dependent motion is extended to parametric representation for two-dimensional time-dependent motion. In fact, careful interpretation of the Euler method to calculate an anti-derivative reveals the accumulation idea underlying the Riemann sum, thus providing a natural introduction to the Fundamental Theorem of Calculus! With only the minor adaptation of setting the initial value of the desired anti-derivative (the "unknown function whose slope is known") equal to zero, students use the same program to calculate Riemann sums to a desired accuracy. (See Figures 3 and 4.)

Clearly, the potential for increasing conceptual understanding (and emphasis upon application) in a calculus course via the availability of a symbolic computation system for illustrating Euler's method and calculating a Riemann sum is exciting; in our course, this has been perhaps the most significant innovation. However, a word of caution is in order regarding students' real understanding of the mathematical processes and techniques involved; this will be considered later.

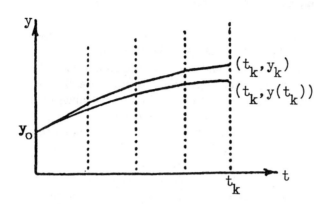

$$\frac{dy}{dt} = f(t,y), \ y(0) = y_0$$
$$\frac{\Delta y}{\Delta t} = f(t_0, y_0)$$
$$\Delta y = f(t_0, y_0)\Delta t$$
$$y_1 = y_0 + \Delta y = y_0 + f(t_0, y_0)\Delta t$$
$$y_k = y_{k-1} + f(t_{k-1}, y_{k-1})\Delta t, \ k = 1, 2, \ldots, n.$$

Figure 3 : The Euler method: numerical solution
of a differential equation.

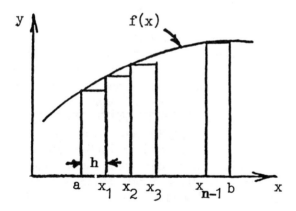

$$y_1 = y_0 + hf(a) = hf(x_0)$$
$$y_2 = y_1 + hf(x_1) = hf(x_0) + hf(x_1)$$
$$\vdots$$
$$y(b) = y(x_n) = y_n = h\sum_{i=1}^{n} f(x_{i-1}).$$

Figure 4 : The Euler method: summation of the
area under the graph of $f(x)$.

2.2 Some Questions Concerning Use of Symbolic Computation

The examples of Part I are intended to represent only a few of the ways in which symbolic computation may be used to enrich the teaching and learning of calculus. The opportunity to use symbolic computation in these contexts (as well as others) has led to some more general questions regarding the uses of symbolic computation in elementary calculus.

2.2.1 What Level of Manipulative Skills Should be Expected of Students?

In 1962, Prof. Weinstock insisted that

> it is useless for a physics student to know the definition of a derivative (as the limit of a certain difference quotient, with an even firmer intuitive grasp of this entity as a slope and as a rate of change) without his being able to perform, with painless facility and complete accuracy, differentiations of reasonably complicated combinations of the basic elementary functions ... [8]

It has been my observation that students who cannot perform such hand calculations with reasonable proficiency and confidence also fail to attain the intuitive grasp of the concept; moreover, they have difficulty interpreting results from machine computation. Ideally, technology and hand calculation should be used together in a mutually supportive environment, firmly based upon understanding the procedures involved.

The first example (the limit of the ratio and the critical point) provides a reasonable backdrop for a closer examination of these fundamental issues. The student may elect, as described in the example, to use the full power of the computation system for exploration and solution of the problem. However, I believe the function $f(x)$ is representative of those that students should be able to treat with hand calculation. Heavy dependency upon a symbolic computation system, unsupported by confidence in one's own ability to perform relatively simple mathematical manipulations, certainly promotes its use as a black box. It fosters yet another type of rote learning—which we are trying to leave behind!

2.2.2 The Euler Method: Is it Easily Accessible for Students?

Not surprisingly, the Euler method (the second example above) provides even greater opportunity for the black box approach. The calculation and plotting of the successive predicted functional values as a solution curve necessitates the ability to write a loop command for the computation system (unless the student is provided with a ready-made program). No doubt the use of some software would have decreased the amount of class time spent on this topic, but it would have also masked the confusion of many students with regard to the algebraic process, the geometric interpretation, or the algorithmic structure. Even after the algorithm was carefully motivated, derived, and demonstrated in class, many students had only a qualitative, and in some cases, an erroneous understanding of the process. We found that it is absolutely essential that several iterations of the algorithm be worked out previously by the student with hand calculations supported by graphical illustrations; only then were students able to truly grasp the method and to form those important linkages with prior mathematical knowledge that are crucial for understanding and for transferability.

Most students were then able, with some guidance on programming the calculator, to formulate and use the relatively simple program needed for the Euler method. We also found this to be an illustration of the maxim offered by Johnson and Riess [3] that "the programming of a numerical method is essential for the enhancement of the concept."

A brief remark about testing may be in order here—one that relates directly to the example of the Euler method. An exam problem requiring use of the tangent-line approximation to predict function values, along with graphical illustration by the student, can elicit much more information about the level of concept mastery than the routine (and somewhat sterile) "Find the equation of the tangent line to the curve..." type problem. Indeed, the two questions address different levels of understanding, and students find the former question more challenging.

2.2.3 Integration: What Should be Emphasized?

No discussion of calculus instruction and the use of symbolic computation systems would be complete without some reflection upon the topic of integration. In many traditional courses in calculus, mastery of the various techniques of integration, with collections of integral problems challenging the ingenuity (or patience) of instructors and students alike, seems to have been a major goal in itself. Machine capability to perform integral calculations (both symbolically and numerically)[2] permits a shift of emphasis away from mechanical aspects of integration towards greater attention to the concept itself.

An ability to formulate a Riemann sum in an applied problem and to express its limiting value as an integral is crucial in many common applications of calculus. As a simple and familiar example from elementary physics, consider the formula for the moment of inertia [6]:

$$I = \int r^2 \, dm.$$

Even assuming that students have mastered the basic idea of the Riemann sum and its limiting form as an integral, i.e., $\sum_i f(q_i)\Delta q_i \rightarrow \int f(q)\, dq$, the leap in generalization and abstraction to the formula is formidable. With less need for emphasizing hand calculation of integrals, we can spend time preparing students for transferring the concept of integration to applied settings and guiding them in understanding the information conveyed by integral forms.

A shift in emphasis, however, must still allow time for the development of reasonable proficiency in manipulation and solution of integrals. The rationale for this has several bases, perhaps the most significant of which relates to the ways integrals are used in subsequent mathematics courses. Integration by parts, for example, is not merely a means of evaluating an integral; it is a tool frequently used in theoretical studies for reducing integrals to more familiar and useful forms. Students should become thoroughly familiar with those procedures. In addition, considerable skill in transforming integrals by some change of variable should be expected. This is often a means for non-dimensionalizing an integral or extracting information about dependence upon parameters; the objective need not be an actual evaluation of the integral. The sometimes tedious manipulations that are involved in mastering the techniques of integration do play a vital role in the development of these skills. It has often been recognized that such practice hones students' algebraic skills, but its importance in developing insight in applied problems and illuminating theoretical results has perhaps not been sufficiently appreciated.

---

[2]At present, symbolic integration on the HP-28S is limited to polynomials.

As an example, consider the formula for the moment of inertia of a sphere of radius R, with constant density ρ, about an axis through the center. Here, using the shell method, an element of mass, dm, is given by

$$dm = \rho dV = 2\pi\rho xy dx = 2\pi\rho x(2\sqrt{R^2 - x^2})dx,$$

so that

$$M = 4\pi\rho \int_0^R x\sqrt{R^2 - x^2}\, dx,$$

whereas

$$I = \int x^2\, dm = 4\pi\rho \int_0^R x^3\sqrt{R^2 - x^2}\, dx.$$

If these integrals are rendered non-dimensional by extracting the parameter R via the substitution $u = x/R$, one arrives at the basic forms

$$I = 4\pi\rho R^5 \int_0^1 u^3\sqrt{1 - u^2}\, du$$

and

$$M = 4\pi\rho R^3 \int_0^1 u\sqrt{1 - u^2}\, du$$

so that

$$I = (\text{quotient of integrals})MR^2.$$

Thus the dependence of I upon characteristics of the sphere, M and R, (up to a constant, to be determined by numerical calculation of the integrals if desired) becomes evident.

The use of this technique to analyze the dependence of I upon characteristics of the sphere becomes even more interesting when the density is not constant; say $\rho = \rho_0 x^\alpha, \alpha > 0$.

Similar analysis then yields

$$\frac{I}{M} = R^2 \frac{\int_0^1 u^{3+\alpha}\sqrt{1 - u^2}\, du}{\int_0^1 u^{1+\alpha}\sqrt{1 - u^2}\, du}$$

or $I = c(\alpha)MR^2$, demonstrating that the moment of inertia still depends upon the same characteristics of the sphere, with a proportionality constant which is now seen to involve only α. A choice of other density distributions (for example, $\rho = \rho_0 e^{-\alpha x}$ or $\rho_0(x^2 + a^2)^n$) reveals a more complicated dependence of I upon M, R, and the new parameters in the density function. The point here is that the demonstration of significant physical principles has been accomplished by transformation of the integrals, with actual

evaluation being only of minor importance. The hand calculation of integrals should no longer be a major goal of calculus instruction, but students must be familiar enough with the techniques of integration to be able to perform some important tasks. They should be able to initiate transformations that render integrals non-dimensional, be able to analyze the role of parameters, and be prepared to extract critical information from the intermediate steps of an integral problem.

2.2.4 A Final Example: Computation vs. Geometrical Insight

A familiar problem in elementary calculus is that of finding the line passing through a given point and just "grazing" a given curve. This is a common application of the tangent line. The solution usually involves some tedious computations using the equation of the line, the equation of the curve, and the derivative. However, certain cases permit alternative (and simpler) means of solution and use methods that may provide valuable geometric or physical insight.

For the case of the circle $x^2 + y^2 = 9$ and the point $(1,5)$, the slope at the "grazing" point (x^*, y^*) is $-x^*/y^*$; this may be substituted into the equation for the tangent line at the point (x^*, y^*). Thus,

$$y^* - 5 = (-x^*/y^*)(x^* - 1),$$

which may be solved simultaneously with the equation of the circle. This might be termed an algebraic, computational approach to the problem. With the availability of a powerful symbolic computation system, this would seem to be a natural approach. The solution of the nonlinear system of equations could be accomplished efficiently, and students would probably not be inclined to further exploration.

Alternatively, this particular problem has an extremely simple trigonometric solution as shown in Figure 5.

This example provides another reminder that we may have to counter a potential overdependence upon computational systems; we must make greater efforts to encourage students to develop flexibility and resourcefulness in solving problems.

3 Finding a Desirable Balance

The examples above have featured elements of modeling, analysis, and interpretation. The calculator or

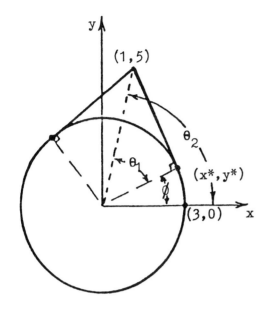

$$\cos\theta_1 = 3/\sqrt{26} \quad \tan\theta_2 = 5$$
$$\Phi = \theta_2 - \theta_1 \quad x^* = 3\cos\Phi \quad y^* = 3\sin\Phi$$

Figure 5: Trigonometric solution.

computer is useful in some instances to deepen and enrich the student's understanding of calculus concepts or applications, or to expedite problem solving. There is no doubt of the great potential of symbolic computation systems for helping students to achieve a successful learning experience in calculus. But in each context I have tried to stress that the student's mastery of the mechanics of calculus is still extremely important. With it, a student remains actively involved with the methods used in problem solving and can intervene, as necessary, using intelligent alternate strategies. Dependence upon symbolic computation must not be so great that students fail to develop abilities to do theoretical work, to recognize relationships and dependencies, and to analyze problem results. Ideally, symbolic computation should complement and extend the students' analytical abilities.

The scientific and engineering communities generally welcome technology that promotes learning and also relieves the burden of tedious manipulation. At the same time, they are concerned with the prospect of students becoming absorbed with this technology, to the possible detriment of their abilities to think critically, to interpret intermediate problem results,

and finally, to judge the reasonableness of answers.

Lastly, although symbolic computation systems permit the solution of more difficult problems, this capability should not be allowed to drive their use. It seems wise to resist temptations to design problems for classroom use whose main virtue is to impress students with the remarkable power of the tool. As far as is possible, students should have a clear understanding of the relevance of examples and problems to course goals. For many students, the line between that which is inspiring and that which is intimidating is finer than we realize.

Considerable progress has already been made in various areas of mathematical research as a direct result of insight provided by symbolic computation. Its impact upon mathematics education and research will be far-reaching. As elementary mathematics courses incorporate various forms of symbolic computation in the future, we should follow the progress of our students in subsequent mathematics and related courses. We must seek a desirable balance between students' development of a full spectrum of mathematical abilities and their dependence upon machine computation. Their capacities for growing mathematically as well as for applying mathematics in their career fields should be our continuing concern.

References

1. Douglas, R. G. (ed.). (1986). *Toward a Lean and Lively Calculus*, MAA Notes number 6, Mathematical Association of America, Washington, D.C.

2. Edwards C. H. Jr. and Penney D. E. (1985). *Elementary Differential Equations with Applications*, Prentice-Hall, Inc., Englewood Cliffs, NJ.

3. Johnson, L. and Riess, D. (To be published). *Calculus with Analytic Geometry*, Harper-Collins, Glenview, IL.

4. Purcell, E. J. and Varberg, D. (1984). *Calculus with Analytic Geometry*, (5th Edition), Prentice-Hall, Englewood Cliffs, NJ.

5. Schrock, C. (1989). "Calculus and Computing: An Exploratory Study to Examine the Effectiveness of Using a Computer Algebra System to Develop Increased Conceptual Understanding in a First Semester Calculus Course," Doctoral Dissertation, Kansas State University.

6. Stanford, A. L. and Tanner, J. M. (1985). *Physics for Students of Science and Engineering,* Academic Press, Orlando, FL.

7. Steen, L. A. (ed.) (1988). *Calculus for a New Century: A Pump, not a Filter,* MAA Notes number 8, Mathematical Association of America, Washington, D.C.

8. Weinstock, R. (1963). "Yes, What Is Happening to Calculus?," *American Journal of Physics,* 31.

9. Zorn, P. (1989). "Algebraic, Graphical, and Numerical Computing in First-Year Calculus," *Proceedings, 1989 Annual Conference, American Society for Engineering Education,* University of Nebraska, Lincoln, NE.

Four Uses of *Derive* in the Instruction of Calculus at the United States Military Academy *

David H. Olwell
United States Military Academy

1 Introduction

The United States Military Academy at West Point has a unique set of resources that favor the success of the inclusion of computer algebra systems (CAS) in the undergraduate curriculum. First, every cadet purchases an Intel 80386-based personal computer (PC) and a personal copy of *Derive*. Second, every faculty member in the Department of Mathematical Sciences has a comparable personal computer and a copy of *Derive*. Third, the department is committed to incorporating CAS into the curriculum, and with a military faculty resistance has been minimal. Through faculty workshops, the department has certified every instructor as proficient in the fundamentals of *Derive*.

The Department of Mathematical Sciences has institutional responsibility to teach every cadet the use of *Derive* (as well as the HP-28S advanced scientific calculator, the *QuattroPro* spreadsheet program, and the MINITAB statistical analysis program). Instructors use *Derive* in the first three courses of the core mathematics curriculum, all of which include projects that require the use of *Derive*. Other departments at the Military Academy, such as physics and the social sciences, are including the use of *Derive* in their instruction. Since the Military Academy has adopted *Derive* as its standard CAS, the Department of Mathematical Sciences has been offering faculty workshops for the other departments on the uses and capabilities of *Derive*. We have also coordinated our instruction on *Derive* with other departmental course material,

particularly physics.

The department is focused on mathematical applications, especially modeling. We view the mathematical modeling process as having three stages. First, a scientific problem is formulated in mathematical terms. Second, the mathematical problem thus created is solved. Third, the mathematical solution is interpreted, empirically verified in the context of the original scientific problem, and the results communicated. We feel that there is a widespread misunderstanding that the second step is the most important and that manipulative skill is the most valued product of the undergraduate mathematics program. We believe the use of *Derive* (and other computational aids) can reduce the amount of time spent by the instructor and student on step two. Students are then able to concentrate on steps one and three, which better engage their curiosity, problem-solving skills, and experimental disposition.

Derive also aids the student's visualization of the calculus. Carefully crafted examples and labs can reinforce the critical concepts of the course by allowing students to see their work graphically.

Our use of *Derive* seeks to capitalize upon these two strengths: we assign tough, open-ended projects that emphasize the modeling process and require the use of *Derive* for efficient completion, and we assign labs and homework sets that require students to construct and graph key results. We are leading students through a four-semester growth process which seeks to produce students who understand viscerally the mathematics they use, and who are not afraid to tackle problems that require them to make assumptions, to construct and solve models, and to communicate the results orally and in writing.

---

*The views expressed herein are those of the author and do not purport to reflect the position of the United States Military Academy, Department of the Army, or the Department of Defense.

This emphasis on problem solving supports the curriculum at West Point, where every student must complete a sequence of core courses that includes four semesters of mathematics, two semesters of physics, two semesters of chemistry, and five semesters of engineering. It also supports the West Point mission of producing technologically competent Army officers.

Derive has been used in student labs, on graded homework, during classroom presentations, on extra credit problems, and with major course projects. In this paper, I would like to share four representative uses of *Derive* in teaching first- and second-semester calculus. The first is a *Derive* lab which serves as my introduction to the first-order differential equations block of the course. The second is a homework sheet given to students during our initial differentiation block of instruction. The third is an in-class demonstration used during my presentation of the double integral. The fourth is a geometric optimization problem that was offered as one of several extra credit projects during Calculus I. These examples are typical of our uses of *Derive* at West Point[1].

I offer some observations about the problems I have encountered incorporating *Derive* into my classes and some strategies for overcoming these problems. The problems are student resistance, the decay of *Derive* skills over time, competition for scarce lab resources, and a paucity of good lab materials.

2   An Introduction to Differential Equations Lab

This lab starts with a review of basic manipulations using *Derive*, and then examines four differential equations and their solutions. I do not introduce any techniques for finding solutions in this lab. I stress the idea of a solution to a DE as a function that satisfies the DE. The calculations involved can be tedious, especially if the students are weak at differentiation and algebra. The use of *Derive* allows the students to easily verify given solutions. The students can then concentrate on the idea of solutions, not on the algebra or differentiation. This lab is conducted in a PC lab. I require the students to finish the lab on their PC's and to submit a lab report with printouts

of equations and graphs.

The lab assignment. In the basic manipulations that are shown below, the boldface type represents the *Derive* commands.

1. **Author** $\exp(2x)$. **Plot.** Find the antiderivative. Find $\int_0^2 e^{2x}\,dx$.

2. **Author** $(x^3 + x^2 + x + 1)/(x(x-2)^2(x+2))$. **Expand** as partial fractions. Find the antiderivative.

3. **Author** $x^3 + 3$. Find the first and second derivatives. Using **Build**, find $y'' + 2y' + y$.

4. **Author** $\sin^2(x)$. **Plot.** **Author** $(1-\cos(2x))/2$. **Plot** both on the same graph. Comment. Set **Manage**, **Trigonometry**, **Collect**, **Cosine**, and then **Simplify** $\sin^2(x)$. Comment.

5. A useful *Derive* function is the CHI function. It is equal to one on an interval, and zero elsewhere. **Author** CHI$(0,x,1)$. **Plot.** **Build** CHI$(0,x,1)$*$\sin^2(x)$. **Plot.**

The first two basic manipulations are a review of previously taught techniques of integration. I include them as reinforcement and to re-familiarize the student with *Derive* using known material. The last three are the *Derive* techniques that we will use in this lesson. In particular, I want the students to be able to test a solution to a differential equation by taking derivatives and substituting them into the DE. Since *Derive* does not automatically simplify all trigonometric expressions, I want them to test for trigonometric identities by comparing graphs. Then I want them to use the **Manage**, **Trigonometry** commands to simplify trigonometric expressions in terms of sine or cosine. Finally, I introduce the CHI function in order to restrict the domain of the solutions to allow for faster computation.

3   Body of the Lab

Problem 1. A differential equation (DE) is an equation involving the derivatives of a function. Without knowing *how* to find solutions (covered in MA205), it is still possible to *verify* solutions to a DE. Example:

DE: $y'' + y = 0$. A solution is $y = \sin(x)$, because $y' = \cos(x)$ and $y'' = -\sin(x)$. Now,

---

[1] The whole-hearted adoption of CAS systems at USMA and the resulting reform of our core mathematics curriculum are the direct result of the vision and leadership of F. Giordano, D. C. Arney, L. Dewald, J. Edwards, R. Kolb, J. Robertson, and R. Schumacher.

$-\sin(x) + \sin(x) = 0$ for all x, so $y = \sin(x)$ is a solution of the DE.

Can you think of another solution and verify it?

Problem 2. Differential equations model a wide variety of physical phenomena. For example $\frac{d^2x}{dt^2} = -32$ models a body falling near the surface of the earth, neglecting air resistance.

1. Verify that $x = -16t^2$ is a solution.

2. Is $x = -16t^2 + 100$ a solution?

3. Is $x = \sin(t)$ a solution?

4. Does it make any sense in these solutions to have negative values of t? Comment.

Problem 3. Including air resistance changes the model of the behavior of the falling body. Assuming the body starts at rest, one such model is: $\frac{dv}{dt} = 32 - 8v$, where v is the velocity which equals $\frac{dx}{dt}$.

1. Verify that $v = 4(1 - e^{-8t})$ is a solution of the DE.

2. If $v = 4(1 - e^{-8t})$ and $x(0) = 0$, find $x(t)$.

3. Describe and plot $v(t)$ and $x(t)$. Do these plots make sense? Does it make any sense to have negative values of t? Multiply $v(t)$ and $x(t)$ by CHI$(0,t,10)$ and **Plot**. Does this make more sense?

4. Rewrite the DE as $\frac{d^2x}{dt^2} = 32 - 8\frac{dx}{dt}$. Verify that $x(t)$ is still a solution.

Problem 4. Another DE, which describes the motion of a damped forced spring, is:

$$\frac{d^2x}{dt^2} + 4\frac{dx}{dt} + 20x = 10\cos(2t).$$

A solution is $x = \frac{-5}{8}e^{-2t}\cos(4t - r) + \frac{\sqrt{5}}{4}\cos(2t - s)$, where $\tan(r) = 34$ and $\tan(s) = 12$.

1. Verify. If you cannot get *Derive* to simplify the left-hand side of the DE to $10\cos(2t)$, try working with the **Manage, Trigonometry** commands to write all trigonometric expressions in terms of cosine. First, convince yourself that you have a solution by **Plotting** the left-hand side and the right-hand side of the DE on the same axes—you should get the same graph each time.

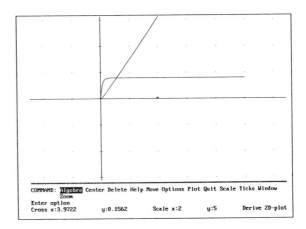

Figure 1: Plots of $v(t)$ and $x(t)$.

2. **Plot** $x(t)$ and $\frac{-5}{8}e^{-2t}$ on the same screen.

3. Interpret.

The last two problems are interesting. *Derive* simplifies the calculations tremendously. In Problem 3, plotting $v(t)$ and $x(t)$ using the CHI function (as suggested in Part 3) allows the student to see that the object reaches terminal velocity ($v_{\text{terminal}} = 4$ units/sec) quickly (see Figure 1). The instructor is then free to follow up with a discussion of falling body behavior, terminal velocity, and a comparison with the earlier model which neglected air resistance. I find it very fruitful to discuss the domains of these two functions and why we exclude negative values of t from this model, and whether we should have an upper limit on t. The student is led to an appreciation of the strengths, weaknesses, and limitations of these simple models.

The damped forced spring problem (Problem 4), is admittedly more difficult, when done by hand. However, it demonstrates dramatically how *Derive* allows the class to ignore the tedium of calculation and get right to the interesting parts: verification and behavior of the solution.

While these commands are straightforward, there is a definite process involved in building student expertise to the level that makes these labs routine. This topic is discussed further in the Section 7. The investment in instructor time and effort is worth the results. It would be very difficult for many of my students to verify by hand that the solution I propose is in fact a solution to the DE. Yet with the CAS, the student can see easily that the solution is valid. The student learns that if one understands conceptually what is to be done, one can be assisted by the CAS to do it. Second, when students encounter difficult problems later in the engineering curriculum, they will have the confidence necessary to outline what manipulations must be made to solve the problem, and then to use the CAS in the actual computations. We are trying to foster an aggressive problem-solving attitude in our cadets. Student who are "power" users of a CAS are well on their way to becoming aggressive problem solver.

Derive offers students and the instructor in this lab a chance to get to the heart of the verification and behavior of solutions of the DE without getting bogged down in the mechanics of computation. This lab allows students to see the how a solution "solves" a DE. they are now motivated to explore the particular DE's covered in this course (linear first-order DE's). A seed for future study has also been planted: other DE's exist, can be solved, and exhibit interesting behavior.

The reports that the students turn in for this assignment are informal. Students submit printouts of their *Derive* work and plots, and is free either to annotate the printouts or respond on separate paper. The reports demonstrate a greater understanding of the relationship between the differential equation and its solution, and an appreciation of the effects of additional assumptions upon model behavior. Prior to the use of *Derive*, the teaching of this material often bogged down in student difficulties with the differentiation and substitution of solutions into the original DE. Now the student can follow the ideas despite any manipulative shortcomings. This has been especially useful when this lab was used with at-risk students who had very weak manipulative skills.

This lab supports our view of the importance of the first and third steps of the modeling process. Students see the problem formulated in mathematical terms and then compares the results of different models with their own empirical experience.

These problems are revisited in the portion of the core curriculum that discusses solutions to second-order differential equations, and the student eventually finds the solutions that are only verified in this lab.

4   A Homework Set

I assign some *Derive* labs as take-home exercises for students to complete in their rooms, with their per-

sonal computers. My objectives for this homework set were to familiarize students with the capability of *Derive* to find derivatives, to reinforce the rules of differentiation, and to reinforce the definition of the derivative as a limit.

Derive **Lab #2: Differentiation.** During this lab, you will use *Derive* to differentiate functions and investigate several of the rules of differentiation. For each exercise, you will be required to answer questions concerning the work performed on *Derive* and the results received. Provide output for each problem. Answers may be written neatly on the output.

1. Find the derivatives of $5x^3 + 17x + \tan(x)$ and $4\csc(x) - 2\sin(x)$.

   (a) What rules of differentiation are applied here? Write them out symbolically.

   (b) Plot each function and its derivative on the same axis.

   (c) What are the domains of the function and the derivative? Comment.

2. Find the derivatives of the $(x^2 + 1)\cot(x)$ and $\sin(x)\sin(x) + \cos(x)\cos(x)$.

   (a) What rules of differentiation are applied here? Write them out symbolically.

   (b) Plot each function and its derivative on the same axis.

   (c) What are the domains of the function and the derivative? Comment.

   (d) Why is the derivative of the second expression zero? Comment.

3. Find the derivatives of $\sin(x)/(x^2 + 5)$, and $(x - 1)/(x + 2)$.

   (a) What rules of differentiation are applied here? Write them out symbolically

   (b) Find the equation of the tangent line to each function at $x = 1$. Plot the function and the tangent line on the same axes. Comment.

4. Find the derivative of $\sqrt{\sin(\tan(x))}$ and $(x^2 - 10x^4 + 4x)^3$.

   (a) What rules of differentiation are being applied here?

   (b) Plot the first function. What is its domain? Explain.

5. Determine the derivatives of the following functions using the definition of the derivative as the limit of a difference quotient (similar to the handout). Evaluate the derivative at the given value, if one is provided.

   (a) $(x+2)^2$,
   (b) $x\sqrt{2x+3}$; $x = 3$,
   (c) $((3x-2)/(2-2x))^{\frac{1}{3}}$,
   (d) $x^2 \sin(\pi x)^3$; $x = 1/6$.

Successful completion of this lab provides the student with a tool to calculate derivatives and with a solid understanding of the meaning of a derivative. The comments on the student report illustrate how *Derive* is aiding student visualization.

5 An In-Class Example

Students often have difficulty visualizing functions of two variables. They also have difficulty constructing double integrals. For example, suppose we are working with the function $f(x,y) = x^2 + y^2 + 10xy$. I first plot the function. Then I ask the students to consider as a region of integration the ellipse described by the equation $x^2 + 2y^2 = 9$. How does the function appear above the region of integration? A simple plot produces Figure 2.

By a sophisticated use of the CHI command, we can view the function above the region of integration, as Figure 3 shows. The equation of the ellipse was manipulated to determine the limits of the CHI statement which are also the limits of integration for the integral. We also ask the students to speculate on the presence and effect of any symmetry in the graphs.

Next, *Derive* is used to evaluate the double integral. After seeing the integration process work, the student is pre-conditioned to listen to the explanation of how these integrals are computed by hand. If one uses a Riemann sum approach when explaining the double integral as a limit process, the rectangular grids of the 3D plot are an extra bonus!

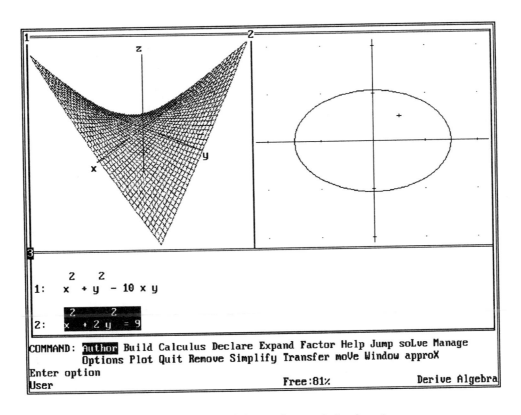

Figure 2: Plot of the surface and the domain.

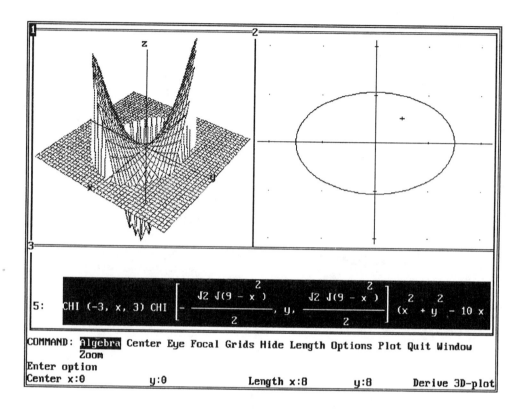

Figure 3: The surface above the domain.

6   An Extra Credit Problem

I use extra-credit problems to allow students to explore some ideas in greater depth. I find that making them voluntary motivates those who attempt them to give their best effort. I grade them liberally. Students find deep satisfaction in the successful completion of these projects. Here is my favorite. It is a modification of a problem I found in Zill [3].

Problem: Given an arbitrary ellipse and any circumscribing isosceles triangle as pictured in Figure 4, prove that the minimal area of the triangle is $3\sqrt{3}ab$ when the altitude is $3a$.

The idea of the solution is simple: the left and right sides of the triangle are tangent to the ellipse at the point of intersection. Position a set of coordinate axes with its origin at the center of the base of the triangle. One finds the side of the triangle as the equation of the tangent line to the ellipse at the point of intersection

and finds the x and y intercepts of that line. The product of the intercepts is the area of the triangle. One then minimizes that area function by the usual calculus means.

What makes the problem difficult is not the concept, but the algebra of rewriting the problem in one variable and then correctly finding the zeroes of the first derivative. Here a clever use of *Derive* is invaluable. I have had students with a correct approach solve the problem in Cartesian coordinates (not my choice) by hand who claim to have spent tens of hours on the problem. They can be shown in class how the same solution can be obtained with a CAS in less than 10 minutes. The steps are the same—only the calculations are faster (and more accurate) by machine. However, they must understand the limitations of the CAS and cooperate by posing the problem effectively and then making informed choices of *Derive* settings. Once they see the ease of this approach, they are invariably hooked on *Derive*. My better students un-

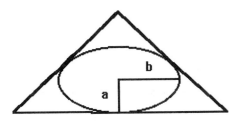

Figure 4: Ellipse with circumscribing triangle.

derstand from the start that *Derive* should be used. For them, the exercise reinforces their use of *Derive* as a tool and polishes their modeling and *Derive* skills.

The reader is invited to try to solve this problem by hand using the techniques of single variable calculus. The manipulations are challenging. However, the level of computational difficulty adds nothing to the student's learning of the calculus. In fact, this problem can do more harm than good when attempted by most students without some CAS. Discouragement sets in and the student's confidence suffers. Yet even weak students can follow the idea of the solution, and then implement it using a CAS.

This problem reinforces for the student that correctly formulating the problem and interpreting the results are as important as solving the mathematical problem. We also reinforce that the conceptual understanding of what is to be done to solve the problem is critical, and that the details of the calculations can be assisted by machine. This problem is revisited in the multivariable portion of the course.

7   Obstacles and Strategies

The adoption of *Derive* and the commitment to incorporate it completely into our instruction has been occasionally difficult for our department and for our students. In this section, I discuss some obstacles to the success of our use of the CAS and some strategies adopted by our department leadership and faculty to overcome the obstacles.

The first hurdle was to ensure instructor proficiency on the CAS. This was fairly straightforward to overcome, because of the unique military composition of our faculty. The faculty works diligently to accomplish department goals, and when we committed to *Derive* we were assured of the enthusiastic cooperation of the faculty. Still, the issue of training the

faculty had to be addressed. The solution adopted was to include *Derive* as a topic in our semiannual Faculty Development Workshops. J. Cummings developed an instructor's guide to *Derive*, which served as a resource for our instructors. Faculty members generously shared their ideas and strategies with each other. This was critical, as there were few commercial resources available. The lack of a comprehensive *Derive* textbook supplement for calculus as a major impediment to success is discussed in greater detail in a later paragraph.

A second obstacle was student resistance to mastering *Derive*. The academic load at West Point is very heavy, and successful students must manage their time well. Cadets realized early on that we would not easily be able to test their mastery of *Derive* on examinations, so many chose to concentrate on other areas. The faculty countered by including questions on the mechanics of *Derive* on some examinations. We then designed academic projects that required the use of *Derive* for efficient solution. We found that we had to constantly reinforce *Derive* skills or they would decay. When the decay occurred, frustrated instructors would bemoan the need to reteach old material.

Our strategy to cope with this was simple: this semester when we designed the syllabus for a course, we identified our *Derive* objectives. We then mapped them onto the syllabus, programming which skill would be taught when, and with what medium. We are currently coordinating this *Derive* "thread" throughout all our courses, identifying the skills that students need to bring forward to successive courses. This overarching design, we hope, will reduce the need to retrain and ensure uniform proficiency in the CAS. We identify the *Derive* goals in the course objectives for the student; thus, there is no uncertainty about the course standard. This helps the students achieve our expectations.

At West Point, we have surprisingly limited computer lab resources because all students have their own PC in their room. This presents a problem: how do we best assist the student to learn *Derive* when we are not physically located with the student's computer? I have had success issuing lab sheets to the students, and then telling them that during the lab hour I will be in my office monitoring electronic mail and the telephone. Students work the lab in their room. As they encounter a problem, they send me a message via electronic mail. I answer it, and if I think it would help, I forward the question and answer to every other student via e-mail. In this instance, we capitalize on our institutional investment in LANs: every cadet PC is hard-wired to the e-mail network. As a second strategy, we have borrowed laboratory resources from the Electrical Engineering/Computer Science Department during hours when they are not in use. Enterprising individual instructors can often make arrangements for their sections, but the high volume of mathematics instruction (over 70% of the students at West Point are taking a mathematics course in a given semester) prevents effective group scheduling.

Finally, we have found the lack of commercial course materials that fully utilize *Derive*, to be an impediment. It takes time for an individual instructor to develop good materials that support the particular lesson objectives for the day's class. It takes even more time to craft *Derive* lab exercises that use the symbolic and graphical capabilities of *Derive* as teaching tools to aid student visualization and comprehension. At least two *Derive* workbooks are expected to appear in print soon (see [1] and [2]). As workbooks become more available, the use of *Derive* in calculus courses should become less labor intensive for instructors.

Such workbooks will still not alleviate the minor obstacle that arises when one considers the amount of time required to grade the many labs and projects in a *Derive*-intensive course. If the instructor does not collect the *Derive* lab work, it may not be completed by the student. Once collected, it takes time to grade it. It is much harder to teach calculus this way than it is to keep to the standard lecture, mid-term examination, then final examination format.

8   Conclusion

The use of *Derive* in the classroom and outside the classroom allows the student to experiment with mathematical concepts and solution techniques without getting hopelessly bogged down in calculations. In each example in this paper, I have tried to show how the ease of computation in a CAS can make difficult problems accessible to average students and allow them to visualize the key calculus concepts. At West Point, we teach students how to use *Derive* to solve significant problems. This builds confidence that they understand the calculus and can work efficiently with it, despite any algebraic or trigonometric weaknesses. They are able to focus significant energy on the first and third steps of our modeling process: constructing

the model and interpreting its solution, while using technology to efficiently complete the computational steps.

At West Point, we are incorporating CAS into all of our core mathematics courses, but particularly into our calculus instruction. With *Derive*, we are producing confident problem-solvers.

References

1. Arney, D. C. (1992). *Exploring Calculus with Derive*, Addison-Wesley, Reading, MA.

2. Olwell, D. H. and Driscoll, P. J. (1992). *Calculus and Derive*, Saunders Publishing, Philadelphia, PA.

3. Zill, D. G. (1992). *Calculus*, 3rd ed., Prindle Weber and Schmidt, Boston, MA, 3–77.

Antidifferentiation and the Definite Integral: Using Computer Algebra to Make a Difference

Phoebe T. Judson
Trinity University

1 Introduction

During the past few years I have encouraged my Calculus I and II students to make full use of the symbol manipulating capabilities of the computer algebra system *Maple*. I am convinced that using *Maple* helps them concentrate on concepts as well as appreciate the usefulness of the calculus, and it prevents them from losing themselves in mindless computation.

In developing laboratory materials, I came across another and perhaps more exciting reason for using computer algebra. By asking students to interpret the output of such a system, I introduce writing into my courses and the written responses from students sometimes show me what they do not understand. This paper describes how I use computer algebra to increase student understanding, particularly about antidifferentiation and definite integration.

We teach mainstream calculus at Trinity University as a standard three-semester course, using conventional texts such as Stein or Swokowski. Calculus I covers limits, continuity, derivatives, antiderivatives, the definite integral and applications of the definite integral. In Calculus II, we introduce the transcendental functions, develop techniques of integration, investigate indeterminate forms and improper integrals, and study sequences and series.

2 The Definite Integral

As we currently teach Calculus I, we introduce antidifferentiation and give the basic properties in one meeting. We then turn our attention to the area under a curve. We begin by approximating the area of a region that lies entirely above the x-axis with lower sums and upper sums, illustrating with inscribed and circumscribed rectangles.

We then assume f is continuous and nonnegative on a closed interval, and we compute, by hand, the area under a few simple curves using partitions with subintervals of equal length, and taking the limit as n approaches infinity of the Riemann sum, using-right hand endpoints. Finally, we define the definite integral. I have found certain *Maple* exercises (some are given later in this section) to be particularly helpful to students. One of the greatest advantages of a computer algebra system is the combination of symbolic, numeric, and graphic capabilities in a single working environment.

For a continuous function, f, it is more convenient to use the following special case of the definition of the definite integral on a closed interval $[a, b]$

$$\int_a^b f(x)\,dx = \lim_{\Delta x \to 0} \sum_{k=1}^n f(w_k)\Delta x.$$

Here $[a, b]$ is partitioned into n equal subintervals, w_k is chosen to be the right endpoint of the kth subinterval, $f(w_k)$ is the height of the rectangle on the kth subinterval, and Δx is the length of each subinterval. To illustrate my approach to teaching the definite integral, I now give an example, exercises, and a discussion of the value of the exercises.

Example. Use the definition of the definite integral to find

$$\int_0^1 (x^2 - 5x)\,dx.$$

Solution: To evaluate this definite integral, using the definition, we first choose n equal subintervals so that

$\Delta x = 1/n$. (Remember, with equal subintervals, Δx is always $(b - a)/n$.)

Choose w_k to be the right endpoint of the kth subinterval, so that $w_k = k*1/n$. The Riemann sum is then

$$\sum_{k=1}^{n} f(k * 1/n)\Delta x.$$

A *Maple* sequence for evaluating this sum is the following:

1. `>with(student);` (Load the student calculus package.)

2. `>f:= proc(x) x^2 - 5*x end;` (Define the function).

3. `>R:= Sum(f(k*(1/n))*(1/n),k=1..n);`
   (*Maple* displays, but does not evaluate the commands in the student package which begin with uppercase letters.) Remember, in this problem Δx is $1/n$ and w_k is $k * 1/n$.

4. `>R:= sum(f(k*(1/n))*(1/n),k=1..n);`
   (*Maple* will evaluate this sum).

5. `>R:= simplify(");` _____.

6. We know that the value of the definite integral can be approximated by this Riemann sum. Substitute the following values for n and report the value of the Riemann sum in the spaces provided. (Aren't you glad you are not doing this by hand?)
   `>subs(n=100,R); evalf(");` (Note, both commands can be written on one line.)
   $n = 100$ _____; $n = 1,000$ _____;
   $n = 100,000$ _____.
   Guess a value for this definite integral:
   _____

7. `>integral:= limit(R,n=infinity);`
   _____.

8. Concluding statement: $\int_0^1 (x^2 - 5x)\,dx) =$
   _____.

9. *Maple* has a built in command for evaluating many definite integrals. You can check your work with these commands:
   `>int(f(x), x=0..1);   evalf(");`
   (Should the answers you get for parts 7 and 8 be identical or just approximately equal? Explain. Attach your explanation to your report.)

10. Illustrate the Riemann sum by drawing a sketch of the integrand, $f(x)$, which includes representative rectangles.
    `>rightbox(f(x), x = 0..1, 6);` will produce a sketch of $f(x)$ on $[0, 1]$, with six rectangles, using the right endpoint of each subinterval to determine the rectangles' heights.

Exercises. Evaluate each of the following definite integrals by following the procedure outlined above, Steps 1 through 9. Fill in all blanks, and write explanations when requested. Caution: exercise care in determining w_k.

1. $\int_0^3 (5x - 7)\,dx$

2. $\int_2^6 (x^3 - x)\,dx$ (Hint: draw a number line to figure out what w_k is.

3. $\int_{-1}^2 (3 - 4x^2)\,dx$

4. Find a general expression for w_k which will work for all intervals $[a, b]$. Show that your expression is valid for problems 1, 2, and 3.
   $w_k =$ _____

5. Use the `int` command to evaluate the integrals in 1 through 3. Explain any differences.

Discussion. This exercise set illustrates an extra dividend provided by computer algebra. Solution of the problems requires students to understand the process, not just mimic a template solution. Students try to do each of the problems by following the example. This technique works perfectly for the first integral, but the second is just a bit different. Even with my cautionary comments, most students blindly follow the example and problem one. The motivation comes when they find out that the result of the `int` command is different from the result of the sum command. This forces them to think through the process.

A few students see that the results are different but do not understand the process well enough to know something is wrong. Some of them decide that *Maple* is failing. They believe the result of the `int` command is wrong because they trust the results of the sum command. A few of them realize that they need to think about w_k a little more. Using computer algebra to do these problems tells me more about student understanding than doing them with paper and pencil. Different results using paper and pencil do not cause students to wonder what is wrong because students assume that they have made a mistake.

These exercises also challenge student beliefs about the limiting process. A well-written response is required to the question posed in part (9) of the example problem: "Should the answers you get for parts 7 and 8 be identical or just approximately equal? Explain." Many students believe the two answers should be "very nearly the same, but not exactly equal." The answers to this question have helped me diagnose student misconceptions concerning limits.

3 Definite Integration and Antidifferentiation

One or two lectures later, the Fundamental Theorem of Calculus allows students a quick and easy way to evaluate definite integrals. I take great care to explain that not all integrands have antiderivatives in closed form and that one cannot appeal to the Fundamental Theorem in all cases. We then study several approximation methods. These lessons are followed by standard applications of the definite integral, as many as time permits before the semester ends. I have developed lab projects to illustrate average value, the mean value theorem for integrals, and area between two curves. I generally assign one or two of these projects.

We begin Calculus II with an introduction to exponential, logarithmic, inverse trigonometric, and hyperbolic functions. We then spend about two weeks on methods of integration. I continue to stress the need for approximation methods, since the Fundamental Theorem cannot always be used. After techniques of integration have been "mastered," and students feel confident in their integration abilities, I give the following *Maple* assignment.

Exercise.

1. Use *Maple* to find $\int_0^1 x \tan(x)\, dx$.

2. Use *Maple* to find $\int x \tan(x)\, dx$.

3. Explain why *Maple* can give us an answer to one but not both of the preceding problems.

Discussion. The first time I gave this assignment, I was surprised by the responses to Question #3. Almost without exception, students wrote: "The reasons *Maple* cannot find an answer for $\int x \tan(x)\, dx$ is that $\tan(x)$ is undefined at odd multiples of $\pi/2$. Since

there are no numbers on the integral sign, *Maple* does not know what the bounds are, and the integrand therefore becomes infinite at every odd multiple of $\pi/2$. *Maple* can do $\int_0^1 x \tan(x)\, dx$ because $\tan(x)$ is not undefined between 0 and 1."

4 Numerical Methods

The responses to the previous exercise indicated to me that my students did not understand that antidifferentiation and definite integration are different operations. This provides me the opportunity to investigate, with considerable class involvement, the difference between $\int_a^b f(x)\, dx$ and $\int f(x)\, dx$. I then assign a project involving numerical methods of integration. I give students a Riemann sum program, a trapezoidal rule program, and a Simpson's rule program, and I ask them to do the following problems.

Exercise.

1. $\int e^{x^2}\, dx$ is an example of an integral without an antiderivative in closed form. Approximate the value of this integral over the closed interval $[-1, 2]$, using each of the three programs.

2. Approximate $\ln(5)$ using each of the three programs, and compare your results with that obtained from using *Maple* or your calculator to evaluate $\ln(5)$.

Discussion. Students had no difficulty doing the first problem. Many were unable to do the second one, however. They had been able to state (and illustrate) the definition of the natural logarithm on a recent exam, but many were unable to make the connection necessary to do this problem, and they tried to integrate $\ln(5)$ from 0 to 1.

The next *Maple* assignment is my last attempt at finding out whether students have understood the processes of antidifferentiation and definite integration.

Exercise.

1. Attempt to find $\int x \tan(x)\, dx$ with pencil and paper. Show every method you try. Write these attempts neatly and explain how you know that you have tried all appropriate methods of integration.

2. Ask *Maple* to find $\int \tan(x)\, dx$.

3. In a previous exercise, many students thought that *Maple* could not find $\int x \tan(x)\, dx$ because

the tangent function is undefined at all odd multiples of $\pi/2$. In the light of *Maple*'s output for
(2), explain why *Maple* cannot find $\int x \tan(x)\,dx$.

4. Explain the difference between an indefinite integral and a definite integral.

Discussion. Many students were able to answer
these questions correctly, indicating that they finally
realized that $x \tan(x)$ does not have an antiderivative
in closed form, and that they now knew the difference
between evaluating definite integrals and finding antiderivatives. A surprising number were still hazy on
the issue.

5   Conclusion

A primary result of using CAS is that it helps pin-
point student misconceptions. By asking students to
explain why *Maple* can or cannot do certain tasks, an
instructor finds out what students are thinking. If a
student cannot do something with pencil and paper,
poor algebraic skills or careless errors are usually assumed to be the cause. If *Maple* cannot do something,
students often come up with bizarre reasons for the
failure. Their reasons sometimes indicate that they
really do not understand a basic concept.

I have been pleased and excited with the results
of my adventures in the use of computer algebra in
teaching Calculus I and Calculus II. It has proved to
be well worth the extra time and effort involved. I
encourage others to also experiment with computer
algebra in their classes. I personally like *Maple*, because students seem to find it so easy to use. I feel
confident, however, that any of these exercises will
work well with any of the symbolic manipulation systems available. The important message is to become
familiar with a system and then experiment.

Lasting Impressions with Calculus Movies *

J. A. Eidswick

California State University, San Bernardino

1 Introduction

When students are all finished with calculus, what do they remember? If you ask, you'll probably get answers like "oh, you know, derivatives and stuff" or "lots of formulas and hard story problems." If you inquire further, you'll find out what you probably suspected, namely that what was learned is of questionable value. For example, "derivatives and stuff" to most students means things like taking the derivative of $\sin(x)$ or the antiderivative of x^3. What do we **want** our students to remember about calculus? What is the bare minimum? Do we want them to be left with a lasting impression of the true meanings of derivative and integral? Certainly we do. Why else would we have them find derivatives from the definition and require them to work problems involving Riemann sums? Unfortunately, our good intentions misfire because students end up with the wrong message. Most see such assignments only as exercises in algebra, the ultimate goal being to get the answers in the back of the book. Later they wonder why we didn't tell them about "the easy way" to begin with. Now, thanks to the wonders of computers, there is a simple way of making the key concepts of calculus stick. The idea is to present them in the form of "movies" or "animations"—very short in duration—but having a lasting effect. Typical initial reactions from today's students are "cool" or "allright"—sometimes they even applaud!—a typical long-term reaction is "I remember those movies you showed us." Why? Because a picture is indeed worth a thousand words and because today's students respect technology, and, yes, because we are serving the instant gratification crowd. Whatever the reason, it works. Of course, remember-

ing that clever movies were shown and remembering their nature are two different things, and many students won't get the real message until they themselves get involved in the process. In Section 4, we consider some activities that involve students in the process. The effectiveness of computer animations in mathematics is well illustrated by the videos in the series *MATHEMATICS!* [1] in which one can, for example, "see" proofs of the Pythagorean theorem. Another nice example is afforded by [3] where one can watch, among other things, a square-wheeled vehicle taking a smooth ride over a rough road. The project *MATHEMATICS!* was funded by ACM/SIGGRAPH (Association of Computing Machinery's Special Interest Group on Computer Graphics) and the movies of [3] were made using *Mathematica* on a Macintosh computer. In this article, we choose a more economical means for making math movies: the HP-48 (HP-48S or HP-48SX) calculator. At the time of this writing, students are much more likely to have access to this machine or its predecessor, the HP-28S, than any other capable hardware/software combination, and the impact is greatest if the movies are shown on the same machine the students carry around with them. Another reason for selecting the HP-48 is that there is something enchanting about being able to watch a movie run in the palm of your hand. Readers interested in making *Mathematica* movies are referred to [5]. Readers interested in making HP-28S movies may obtain the necessary information by contacting the author. The HP-48 comes with 32K useable memory, which is adequate for the kind of math movies discussed in this article. HP-48SX users interested in making more elaborate HP-48 movies may do so by expanding their useable memory to up to 544K via expansion cards (X stands for expandability). How does one make movies with a computer? The idea is similar

---

*The idea of making movies on the HP-28S was suggested by Mel Thornton

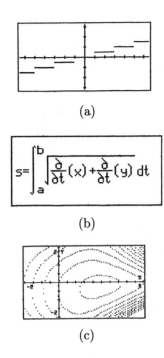

(a)

(b)

(c)

Figure 1: HP-48 Pictures.

to that of making animated cartoons. First you make a collection of pictures, or frames; then you put them together in rapid succession. All that is needed is a way to store and retrieve pictures plus a way to control the time between successive frames. Unlike animated cartoons, mathematics movies lose little if the motion does not flow continuously, and, consequently, relatively few frames are needed to make an effective mathematics movie. There are two challenges: One is to make the pictures; the other is to put life into them.

2   Making Pictures

Calculator/computer displays are really grids that contain a large number of tiny squares or **pixels**. Each pixel can be either shaded (on) or unshaded (off). The size of an HP-48 pixel is roughly 0.47 mm × 0.47 mm, and, in terms of pixels, the display dimensions are 131×64. Given good lighting, the proper angle, and reasonably good eyesight, these pixels can be seen with the naked eye. One way to see them clearly is to look closely at the screen image projected by a classroom display unit. By a **picture** is meant any of the $2^{131 \times 64}$ possible shaded pixel configurations. See Figure 1.

The simplest and most useful tools for making mathematical pictures are the commands DRAW and PIXON. (It is of interest that the calculator can also be used to draw pictures "by hand" like one would draw on an etch-a-sketch™; however, that capability will not be a concern of this article.) The DRAW method, described in the HP manuals, is for graphing functions and is very simple: enter the function of your choice, then press (the white keys under) $\boxed{\text{STEQ}}$ and $\boxed{\text{DRAW}}$ (in that order) in the PLOTR subdirectory of the PLOT directory. For all practical purposes, the DRAW method can be used to graph **any** function—even discontinuous ones. For example, to graph the step function in Figure 1(a), key in the following:

> ≪   −3   3   FOR   I   X   I   2   *   1   +
> <   I   2   /   NEXT   0   −3   3
>
> FOR   I   IFTE   NEXT $\boxed{\text{ENTER}}$
>
> $\boxed{\text{EVAL}}$ $\boxed{\text{STEQ}}$ $\boxed{\text{DRAW}}$

Note that if the HP-48 is in connect mode (CNCT), it will try to "fix" jump discontinuities by automatically inserting steep line segments at those points. This feature can be disabled by pressing the connect key in the MODES menu. In this article boxed-in words (e.g., $\boxed{\text{ENTER}}$) represent single keystroke operations. Other commands like IFTE (if-then-else) may be entered either letter-by-letter or as a single keystroke from an appropriate directory. For a complete list of commands and other operations as well as the directories that contain them, see the HP manuals. For an introduction to the nuts and bolts of the HP-48, see Chapter 0 of [4]. The PIXON method is based on the fact that the sequence of keystrokes $(x \quad y)$ $\boxed{\text{PIXON}}$ produces a dot at the point (x, y). The implied coordinate system here has its origin at the center of the display. From the upper left-hand corner of the display, the origin is exactly 33 pixels down and 66 pixels to the right. The default unit of length corresponds to ten pixels; thus, the standard viewing window is the rectangle $[-6.5, 6.5] \times [-3.1, 3.2]$. For additional details concerning plot parameters, rescaling, etc., see the HP manuals and [4].

Examples.

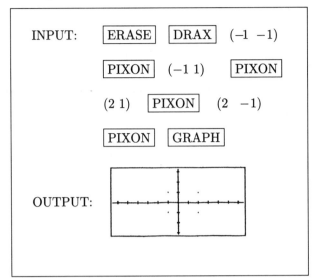

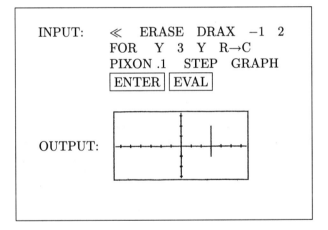

3 Putting Life into Pictures

This can be done by a 3-step process: (1) encode the pictures to be used, (2) assign names to the codes, and (3) use names in a program like the one given below. In the language of the HP-48 Owner's Manual, the current display can be represented by a string, then stored as a global variable which can be used as a procedure object in a program.

To encode a picture that is on the display, press STO, then ON. To encode a picture produced inside of a program, insert PICT RCL at the end of the program. In either case, the desired code will appear on level 1 of the stack in the form "Graphic $m \times n$." To assign a name NAME to a code on level 1 of the stack, type NAME STO; the word NAME will immediately appear in the VAR menu. To get the code

back, press the white key under NAME; to get the picture back, press PICT STO GRAPH.

Examples.

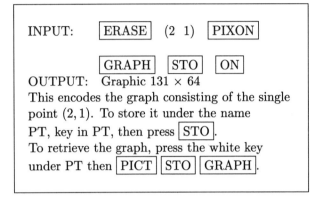

OUTPUT: Graphic 131×64
This encodes the graph consisting of the single point $(2, 1)$. To store it under the name PT, key in PT, then press STO.
To retrieve the graph, press the white key under PT then PICT STO GRAPH.

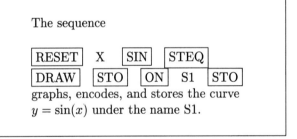

The sequence

RESET X SIN STEQ
DRAW STO ON S1 STO
graphs, encodes, and stores the curve $y = \sin(x)$ under the name S1.

For the movie programs discussed in this article, we will make use of the following auxiliary programs to access the graphics environment and to manipulate graphics objects. Readers intending to enter the programs contained in this paper on their calculators are advised to keep them all together in a single directory. To open a directory called MOVIES (say), key in MOVIES then press CRDIR in the MEMORY directory. The names of the programs below are a hint as to what they do.

```
≪ {#0 #0} PVIEW ENTER G.E.
  STO
≪ PICT {#0 #0} ROT REPL
  ENTER SEE STO
```

Suppose now that we have stored a sequence of displays S1, S2, ..., SN and want to put them together in that order to make a movie. The following program, called MOVIE, does the job.

```
≪  →  N  W  T  ≪  G.E.  1  T
   FOR  J  1  N  FOR  I  "S"  I
   +  OBJ→  SEE  W  WAIT  NEXT
   NEXT  ENTER  MOVIE  STO
```

The word MOVIE will appear in the VAR menu. To run it, type: N W T MOVIE where N is the number of frames, W is the desired time between frames in seconds, and T is the number of times you want to run the movie.

Example 1. Encode and store the curves $y = \sin(x - \pi(n-1)/2)$, $n = 1, 2, 3, 4$, under the names S1, S2, S3, S4, respectively. Then run MOVIE with N = 4, W = .2, and T = 10: 4 .2 10 MOVIE. The result is a sin wave that marches steadily to the right.

Example 2. In this example (also discussed in the HP-48 Owner's Manual), we store $\sin(x)$ and the first nine distinct partial sums of its Maclaurin series under the names S1, ..., S10 to make a movie illustrating convergence. (The version in the HP manual suffers from jerkiness caused by mixing the graphics environment command ERASE too closely with the stack environment command →LCD.) The following program when run will encode the frames.

```
≪  ' PPAR '  PURGE  ERASE  DRAX
   X  SIN  STEQ  DRAW  PICT  RCL
   ' S1 '  STO  1  9  FOR  I  ERASE
   DRAX  X  SIN  X  2  I  *  1  −
   TAYLR  STEQ  DRAW  PICT  RCL
   S1  +  " ' S ' " I  1  +  +  OBJ→
   STO  NEXT  ENTER  MSTR  STO
```

To run the program, press MSTR. Note that to store MSTR takes a very small amount of user memory (about 0.15 Kbytes); to run it takes about 11 Kbytes. (Generally, it takes about 1.1 Kbytes per frame. You can check the amount of available memory at any time by pressing MEM in the MEMORY directory.) To view the movie, enter 10 1 1 MOVIE. When you are done, you will want to purge the graphics strings S1, ..., S10 from user memory so that you can free up those 11 Kbytes. (To regenerate the strings, just press MSTR.) To purge the strings, enter {S1 S2 S3 ... S10} PURGE. (Note that you can

enter S1, ..., S10 from the VAR menu by pressing the white keys under those names.)

4 Student Activities

The simplest way to get students to focus on the mathematical message of movies like Example 2 is to ask them to explain the message after they see the movie. One can do this by holding a classroom discussion about it or by having students write a paragraph or two about their understanding. Another way to get students actively involved is to invite them to participate in the movie-making business. Many students will be receptive to this idea—especially those already intrigued with computer technology. For the necessary background material, Sections 2 and 3 above is more than enough. For student movie-making projects, the programs G.E., SEE, and MOVIE can be taken as black boxes without losing anything. Of course, student movie-makers need to be told to explain their work, especially the mathematical aspects of it. Successful projects can be shared with other members of the class either by passing around the movie or by showing it using a classroom projection device.

Project 1. Make a movie showing a tangent line moving along an interesting curve. Describe the action on a section of the curve that is (a) concave upward; (b) concave downward. Describe what happens near an inflection point. Part of the problem is to figure out what is interesting and what is not. Students who have studied polar or parametric representations will have the potential to do more.

Project 2. Make a movie showing the circle of curvature rolling along an interesting curve. Solving this problem requires and enhances understanding of curves, tangents, normals, and curvature. Here, the selection of an interesting curve is challenging because most familiar curves have flat spots (which are uninteresting). An example of an interesting curve is a cardioid.

Project 3. Make a movie showing a cycloid being generated by a rolling circle. To complete this project requires and enhances understanding of cycloid, arc length, and parametric representation; it also requires some HP graphing skills to construct the appropriate frames. Similar projects may be assigned concerning hypocycloids, epicycloids, and, generally, trochoids. For a discussion of such curves, see [2].

5 Derivative and Integral Movies

The two examples of this section are somewhat more elaborate than those discussed above. The idea here is to illustrate the geometric meanings of derivative and integral in a short period of time. Each movie takes about 40 seconds to show. The programs contain certain embellishments (blinkers, displayed messages, live graphing and variations in speed) that call for a more sophisticated version of the program MOVIE that was previously introduced. Abbreviated outputs of these programs are shown in Figures 2 and 3. The code that produces these figures is somewhat lengthy and is not given here. Readers who wish to obtain the code may do so by contacting the author. Those who do not want to go to the trouble of keying in the programs may make arrangements for a direct transfer of the programs.

References

1. Apostol, T. M. (1988). *Project MATHEMATICS!*, Mathematical Association of America.

2. Hall, L. M. *Trochoids, Roses, and Thorns— Beyond the Spirograph*, to appear in *College Math. J.*

3. Hall L. M. and Wagon, S. *Mathematical Roads and Wheels*, to appear in *Math. Magazine.*

4. Mathews, J. and Eidswick, J. (1992). *The HP-48/28 Calculus Companion*, Harper Collins Publishers.

5. Wagon, S. (1991). *Mathematica in Action*, W.H. Freeman, New York.

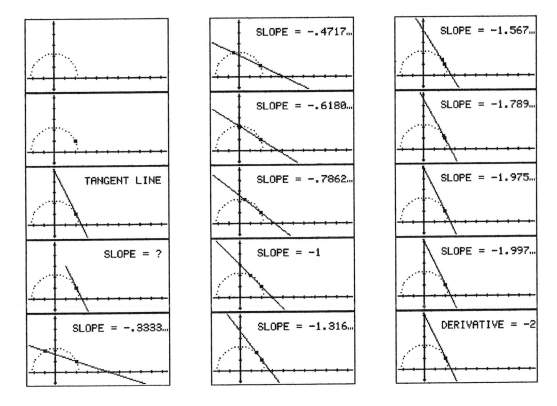

Figure 2: Derivative movie frames.

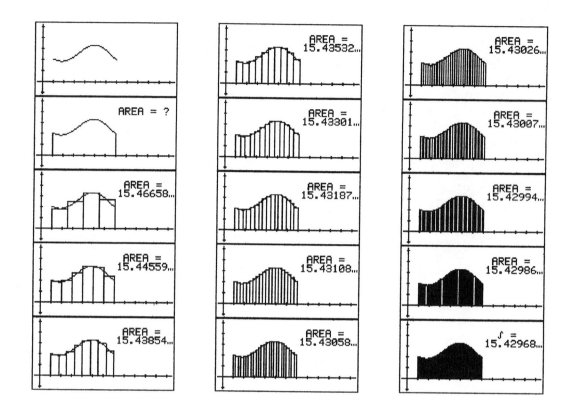

Figure 3: Integral movie frames.

Part III:

Symbolic Computation in
Linear Algebra and Differential Equations

Some Examples of the Use of a CAS in Teaching Linear Algebra and Calculus *

J. W. Auer
Brock University

E. R. Muller
Brock University

1 Introduction

Experiments with computer algebra systems (CASs) in undergraduate courses are now being reported regularly. So far these have tended to be for small classes, and/or for a single section of a multi-section course, and the emphasis has usually been on calculus. At Brock University, CAS laboratories are now firmly established as a component of large-enrollment service courses in calculus, and each year some 800 students take these compulsory laboratories. Muller [6] has outlined the sequential development of these laboratories, described their aims and given examples of the student activities. Auer (see [1], [2]) has recently introduced CASs into first-year linear algebra and has developed specific laboratory activities for that purpose. This paper provides several examples of activities designed for student use of a CAS in linear algebra and one example of a more exploratory calculus activity.

In teaching and learning introductory linear algebra, the greatest benefits CASs appear to be:

1. a comfortable environment for doing the many repetitive numerical calculations that are inherent in linear algebra problems. Students can concentrate on the concepts, while the machine looks after the arithmetic.

2. rational arithmetic, an enormous benefit in elementary linear algebra.

3. user friendliness, very little programming is required; many CAS operations have the same names as the standard mathematical operations that they implement. Our experience has been that most students quickly learn the necessary procedures, and that the laboratories help them to become familiar with the standard terminology of the subject. In fact, the CAS seems to encourage students to use the terminology precisely.

4. symbolic manipulation, which allows students to examine without drudgery numerous instances of general properties and theorems, and also to avoid the frustration caused by algebraic errors.

Similar remarks apply to the teaching and learning of calculus. For that purpose, CASs are used principally to explore the qualitative behavior of functions, their derivatives, and their antiderivatives. Some capabilities provided by CASs in this context are:

1. limits of functions and sequences;

2. plotting of functions, part of the visualization of mathematics, see Tall [9];

3. derivatives, antiderivatives, and differential equations;

4. a comfortable environment, both for numerical computations (approximations) and for algebraic manipulations (especially simplifications).

While it is too early to say that the way we use CASs in mathematics teaching and the kinds of student activities we generate are unqualified "good

---

*The authors wish to thank Professor J. P. Mayberry for his valuable assistance and comments on this paper; and also the editors for their constructive suggestions which have improved the readability of the paper.

things," it is already clear that CASs cannot be ignored by mathematics educators: they certainly offer a new paradigm beyond paper, pencil, and calculator. According to Hillel [3], "CASs can be used as a computational tool or also as an investigative tool ... to put it a little differently, one mode aims to generate results and the other mode aims to generate ideas." He further proposes that CASs are learning tools where "the notion of a learning tool is predicated on a didactical bias that conceptual understanding is best achieved through students' direct 'hands-on' experience which is afforded by interactions with a computer." In some sense the term "tool" understates the role that CASs can perform in the teaching and learning of mathematics. CASs provide a natural and comfortable environment for doing mathematics. One is tempted to think of them as mathematical hypermedia. Some of the philosophical and pedagogical issues are explored in the articles by W. Page [7], and by Hodgson and Muller [4], cited in the References.

It would appear that, as of this writing, the two principal CASs that provide all the benefits noted above are *Mathematica* and *Maple*. Although all our examples did in fact use *Maple* software, *Mathematica* provides analogous capabilities; indeed, the *Mathematica* printouts would appear quite similar.

2   Use of CASs in Teaching Linear Algebra

Some of the typical introductory linear algebra problems for which a CAS is a useful tool are:

> vector algebra calculations
> vector identities
> solution of systems of linear equations
> Gauss-Jordan reduction
> matrix algebra
> inverse matrix problems
> Jacobi and Gauss-Seidel methods
> determinants, cofactors
> checking axioms for vector spaces
> basis, rowspace calculations
> linear transformations
> eigenvalue, eigenvector calculations
> characteristic and minimal polynomials
> power method, deflation
> checking axioms for inner products
> Gram-Schmidt calculations
> complex linear algebra calculations
> simplex method

The following are a few sample elementary linear algebra problems which we feel are especially enhanced by the use of CASs. Each problem statement is followed by the *Maple* session used to solve it. The only user input is on lines beginning with the *Maple* prompt > or a secondary prompt ≫. All the rest (except for comments and documentation appearing in italics) is output from the computer. (For brevity, much of the output of the example that follow has been suppressed.) The amount of typing required of the user is very small, and can be further reduced by using "copy and paste" commands (not described below) found in the Macintosh environment. Note that *Maple* interprets the usual double quotation mark ("), when entered by the user, as the entire result of the preceding action.

2.1   Verifying the Polarization Identity

Because of the symbolic manipulation capabilities of CASs, students are encouraged to explore theorems and identities using their enhanced ability to work out complex expressions involving variable vectors. We have found that the use of a CAS has highlighted the difference between scalar and vector operations. At each stage the student is required to decide which type of operation is appropriate: e.g., whether the * or the `dotprod` operation should be used. Some students forget that $(1/4)|\,\vec{u} + \vec{v}\,|^2$ is ordinary multiplication of numbers; *Maple* reminds them by refusing to do a scalar multiplication! It should be remarked that there are, of course, other ways to present this problem, with a view to stimulating "discovery" of the identity. For example: students, given only the right-hand side, could be asked to identify it, or to find a simpler expression for it.

Problem: Use *Maple* to explore the polarization identity in $\Re^3$:

$$\vec{u} \bullet \vec{v} = (1/4)|\,\vec{u} + \vec{v}\,|^2 - (1/4)|\,\vec{u} - \vec{v}\,|^2,$$

where $\vec{u}$ and $\vec{v}$ are vectors in $\Re^3$.

We start a *Maple* session for this exploration by using the `array(1..n);` command to define $\vec{u}$ and $\vec{v}$ as vectors with n entries. Next, we use the commands `dotprod(u,v);` (for the dotproduct of $\vec{u}$ and $\vec{v}$), `add(u,v);` (for adding vectors $\vec{u}$ and $\vec{v}$), `scalarmul(u,c);` (for the scalar multiple of the vector $\vec{u}$ and scalar c) to obtain **rhs** and **lhs**, the right- and left-hand sides, respectively, of the given identity.

A simple comparison of `rhs` and `lhs` at this point established the identity. (In order to give the result the mnemonic name u+v and u−v must be enclosed in backward quotes "'.")

```
> u:=array(1..3); v:=array(1..3);
        u := array ( 1 .. 3, [])
        v := array ( 1 .. 3, [])
> lhs:=dotprod(u,v);
> 'u+v':=add(u,v);
> '|u+v|^2':=dotprod('u+v','u+v');
> 'u-v':=add(u,scalarmul(v,-1));
> '|u-v|^2':=dotprod('u-v','u-v');
> rhs:=(1/4)*'|u+v|^2'-(1/4)*'|u-v|^2';
> expand(rhs);
        u[1] v[1] + u[2] v[2] + u[3] v[3]
```

2.2 Eigenvalues and Eigenvectors

This is an area of linear algebra which is also of great interest to many scientists. Our colleagues from the sciences would certainly be pleased if we could teach students the concepts while providing them with an environment in which they do not get dragged down in a numerical bog. *Maple* and other CASs provide this environment: the following session takes under three minutes, while a manual solution would take a student close to an hour. We provide this example to counter the criticism that CASs are black boxes from which one does not learn the mathematical concepts. *Maple* can, of course, be used as a black box. For example, it can provide solutions for systems of linear equations, and eigenvalues of matrices with either numeric or symbolic coefficients. But, *Maple* also contains operations that allow step-by-step procedures, and these are valuable from a pedagogical point of view. In what follows we show how a student can use a CAS to solve an eigenvalue problem, much as an instructor would give an example in a lecture and a student would do by hand. Note the benefits of rational arithmetic and the way in which *Maple* removes the drudgery (and the chance of error) without concealing either the concepts or the arithmetic details. In particular, students must be aware of all the concepts and procedures: *Maple* only does the arithmetic and bookkeeping.

Problem: Find the characteristic polynomial, eigenvalues, and eigenvectors of

$$\begin{bmatrix} 5 & -4 & -1 & 3 \\ 3 & -2 & 4 & -5 \\ 0 & 0 & 1 & 1 \\ 0 & 0 & 1 & 1 \end{bmatrix}.$$

Using *Maple*, we first define the matrix given in the problem (through the `array` command) and start constructing the matrix $F = A - xI$. This is done by using the `band` command to obtain the 4×4 matrix $-xI$, and adding (the `add` command) this to A. Now, $\det(F)$ can be obtained (through `f := det(F);`) and solved for x (`soln := solve(f,x);`) to determine the eigenvalues, 0, 1, 2, and 2, of A.

To find the eigenvectors corresponding to the eigenvalue 0, x is set to 0, (`x := soln[1];` actually sets x to the first eigenvalue). (If we wish, we can take a look at the matrix F with x set to 0 by simply using `print(F);`.) The command `kernel(F)` produces a set of basis vectors ([22, 5/2, -1, 1] in this case) for the null space of the transformation defined by F. Obviously, the last part of our *Maple* interaction can now be repeated with x set to the other eigenvalues of A.

```
> A := array([ [5,-4,-1,3], [3,-2,4,-5],
           [0,0,1,1], [0,0,1,1] ]);
> X := band([-x],4);
> F := add(A,X);
> f := det(F);
                        2     3     4
      f := - 4 x + 8 x  - 5 x  + x
> solns := solve(f,x);
          solns := 0, 1, 2, 2
> factor(f);
> eigenvals(A);
> x := solns[1];
> print(F);
> kernel(F);
          [22, 57/2, -1, 1]
```

2.3 Determinants and Cofactors

Here we present two examples. The first example always impresses students and is a simple demonstration of symbolic manipulation with *Maple* which can easily be duplicated using pencil and paper. Once again, we feel this algebraic power of the CAS will encourage students to explore ideas. The second example is certainly nontrivial using only pencil and paper and would tax the best and most careful student in the class. The bookkeeping task is enormous. What is particularly striking is *Maple*'s ability to arrange terms in a way in which the expansion of the determinant can be recognized.

Problem: Show that for the matrix

$$A = \begin{bmatrix} 1 & a & a^2 \\ 1 & b & b^2 \\ 1 & c & c^2 \end{bmatrix}$$

$\det(A) = (b-a)(c-a)(c-b)$.

In our *Maple* session we first define A then obtain $\det(A)$. It is unlikely that the expression obtained for $\det(A)$ will be identical to $(b-a)(c-a)(c-b)$. However, if factored (the `factor("")` command), this expression will become $(b-a)(c-a)(c-b)$. That's all there is to it!

```
> A:=array([[1,a,a^2],[1,b,b^2],[1,c,c^2]]);
> det(A);
> factor(");
```

Problem: Explore the relation between $\text{adj}(\text{adj}(A))$ and A.

Using pencil and paper students can quickly establish that, for a 2×2 matrix A, $\text{adj}(\text{adj}(A)) = A$. It does not take long for them to realize that, for A a 3×3 matrix, the bookkeeping of the various terms is virtually out of hand. However, understanding the linear algebra concepts and using a CAS, students are led to conjecture the identity $\text{adj}(\text{adj}(A)) = (\det(A))^{n-2}A$ when A is $n \times n$, $n \geq 2$, a conjecture that is not at all apparent from a study of the 2×2 case alone. In the sample *Maple* interaction given below, the command `adj(A)` produces the classical adjoint matrix of A. When `B := adj(adj(A))` is used, the output fills several pages. From the experience gained by doing the previous problem, students know that factoring the entries of B **may** produce a pattern. Two "for loops" are used to factor each entry B (The `do` - `od` statements "bracket" the beginning and end of the loops similar to "begin - end" statements in Pascal.) In the *Maple* interaction given below, a portion of the output for B (before as well as after factorization) is included to give the reader a sense of what the student sees as this analysis progresses. It is clear from the last output that all $B(i,j)$ are products of $A(i,j)$ and a common factor. Once it is observed that this common factor is $-\det(A)$, it becomes clear that $\text{adj}(\text{adj}(A)) = \det(A)A$. (Remark: In tests of several other computer algebra systems, the case $n \geq 4$ caused system crashes, thereby illustrating the present constraints imposed on symbolic computation by computer memory and speed.)

```
> A:=array(1..3,1..3);
> B:=adj(adj(A));
B := array ( 1 .. 3, 1 .. 3,

                2
[A[1, 1]  A[3, 3] A[2, 2] - A[1, 1] A[3, 3]
  A[1, 2] A[2, 1] - A[1, 3] A[3, 1] A[1, 1]
                2
 A[2, 2] - A[1, 1]  A[2, 3] A[3, 2] +
 A[1, 1] A[2, 3] A[1, 2] A[3, 1] + A[1, 3]
 A[2, 1] A[1, 1] A[3, 2],

    ... remaining output suppressed
```

```
> for i to 3 do
>>   for j to 3 do
>>   B[i,j]:=factor(B[i,j])
>>   od;od;
```

```
B[1, 1] :=

  - A[1, 1]
    (A[2, 1] A[3, 3] A[1, 2] - A[2, 3]
    A[3, 1] A[1, 2] - A[1, 3] A[3, 2]
    A[2, 1] + A[2, 2] A[3, 1] A[1, 3]
    - A[1, 1] A[2, 2] A[3, 3] + A[1, 1]
    A[3, 2] A[2, 3])

B[1, 2] :=

  - A[1, 2]
    (A[2, 1] A[3, 3] A[1, 2] - A[2, 3]
    A[3, 1] A[1, 2] - A[1, 3] A[3, 2]
    A[2, 1] + A[2, 2] A[3, 1] A[1, 3]
    - A[1, 1] A[2, 2] A[3, 3] +
    A[1, 1] A[3, 2] A[2, 3])

    ... remaining output suppressed
```

2.4 Checking Inner Product Axioms

The classification of inner products is one of many important problems of linear algebra, and the verifying of axioms is a big and traditionally intimidating task for students. This example demonstrates the ease with which students can use *Maple* for axiom verification. As noted earlier, the power of *Maple* builds students' confidence. We feel this greatly improves the learning process.

Problem: Determine whether the rule $< \vec{u}, \vec{v} >= u_2v_2 + 2u_1v_1 + u_1v_2 - u_2v_1$ defines an inner product on $\Re^2$.

Using *Maple*, students first define the vectors $\vec{u}_1$, $\vec{u}_2$, $\vec{v}_1$, and $\vec{v}_2$. Then the definition of $< \vec{u}, \vec{v} >$, given in the problem is established (under the name prod as a *Maple* procedure through the proc ... end statement). Now students begin to check if prod satisfies the four properties required of an inner product. Simplification of the result (through the normal command) of prod(add(u1,u2),v1) - prod(u1,v1) -prod(u2,v1) yields zero, establishing one of the properties of the inner product. Similarly, when the result of prod(scalarmul(u1,c),v1) - c*prod(u1,v1) is simplified we get zero, establishing another property of the inner product. Next, prod(u1,v1) - prod(v1,u1) is checked, yielding
$$2\ \mathtt{u1[1]}\ \mathtt{v1[2]} - 2\ \mathtt{u1[2]}\ \mathtt{v1[1]}.$$
Since this is not zero for certain choices of vector components, we see that this property fails and the given $< \vec{u}, \vec{v} >$ is not an inner product. Students may now wish to check the last property of inner products through prod(u1,u1) to find out that it holds.

```
> u1:=array(1..2); u2:=array(1..2);
> v1:=array(1..2); v2:=array(1..2);
> prod:=proc(u,v) u[2]*v[2]+2*u[1]*v[1] +
        u[1]*v[2] - u[2]*v[1]; end;
> prod(add(u1,u2),v1)
        -prod(u1,v1)-prod(u2,v1);
> normal(");
                0
> prod(scalarmul(u1,c),v1)-c*prod(u1,v1);
> normal(");
                0
> prod(u1,v1)-prod(v1,u1);
        2 u1[1] v1[2] - 2 u1[2] v1[1]
> prod(u1,u1);
                2            2
        u1[2]    + 2 u1[1]
```

3 Use of CASs in Teaching Calculus

In changing our area of interest from linear algebra to calculus we move to one where we have had more experience using *Maple* and therefore more time to reflect on calculus student activities and on in-class presentations. The former are far more developed

than the latter. What is clear is that whereas students are prepared to explore ideas in other disciplines, they have no capacity to explore mathematics concepts. They are driven to look for a single right answer. Our in-class teaching must therefore incorporate exploration activities that we will expect of our students. If we want our students to go through the following example on their own or in a formal laboratory period, we need to spend the time in class doing similar activities. The faculty will find the in-class dynamics very different (so will the students) and a discussion on how to take notes during such activities will not be a waste of time.

We now consider the example and a demonstration of how the rich CAS environment enhances the teaching and learning of the mean value theorem (MVT). When we restrict calculus to paper-and-pencil calculations we find that the functions to which the MVT can be applied are very limiting. There are a number of reasons for this. From a visual point of view it takes the average student a long time to graph the functions and thereby see what is going on. This limitation does not allow for the interplay between the graph of the function and the graph of its derivative, a capability which provides a real revelation for those who are visually oriented. From a numerical point of view, the student's inability to solve the equation $f'(c) =$ "slope of chord" will limit the scope of the exploration to the simplest of functions. The following exercise illustrates the advantage of working with a CAS. It is clear that the exercise need not be limited to especially simple functions.

Problem:

1. Look up the mean value theorem (MVT) in a calculus text and state it in your own words. Note that the theorem provides no clue as to how to obtain the point(s) but only guarantees their existence.

2. You are provided with the function
$$f(x) = \frac{1}{\sqrt{(x^2 + 2x - 3)}}.$$

   (a) Explain why the MVT does not hold in the interval [0.5,1.5].

   (b) Find the point(s) which satisfy the MVT in the interval [1.1,2.2], find the equation of the tangent line(s) at this (these) point(s) and graph them together with the function.

(c) Using $f(x)$ and the interval $[1.1, 2.2]$ explore the following proof techniques for the MVT. Define secant and $e(x)$ by

$$\text{secant} = (f(b) - f(a)) / (b - a),$$

$$e(x) = f(x) - (f(a) + \text{secant} \times (x - a)).$$

(It is interesting to note that $e(x)$ in fact determines the error we would commit should we decide to approximate the given function $f(x)$ by the straight line $f(a) + \text{secant} \times (x - a)$.) Since $e(b) = e(a) = 0$, and $e(x)$ is not equal to 0 for all x in (a, b), $e(x)$ must have a maximum or minimum in that interval: in other words, point(s) where $e'(x) = 0$. Differentiating $e(x)$ and equating to zero we find the statement of the MVT.

3. Apply the MVT to a parabola expressed in general form.

This is how the *Maple* session proceeded for Parts 2(a), 2(b), 2(c) and 3.

Part 2(a): By the definition of $y = f(x)$, a graph of $f(x)$ is obtained for $x \leq 0.5 \leq 1.5$. This graph shows that the MVT cannot be applied on $[0.5, 1.5]$.

```
> y:=(x^2+2*x-3)^(-1/2);
> plot(y,0.5..1.5);
```

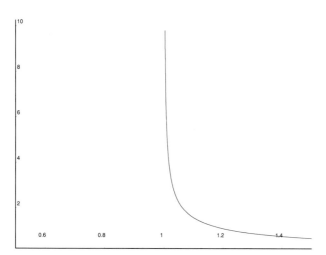

Part 2(b): $f(x)$ is considered on $[1.1, 2.2]$ and a graph of the function is obtained. Since this does not seem to pose any problems, the slope of the line joining $(1.1, f(1.1))$ and $(2.2, f(2.2))$ is calculated.

In the *Maple* expression for slope, $\text{subs}(x=2.2, y)$ gives the result of substituting 2.2 for x in y. Next, the value of c_1, prescribed by the MVT is computed through the `fsolve` command. Here $\text{diff}(y,x)$ is the derivative of y with respect to x; $\text{fsolve}(z,x,a..b)$ solves the equation $z = 0$ for x in the range $[a, b]$, using floating point arithmetic. Now the equation of the tangent line can be obtained (this portion of the output is included in the interaction given below) and then graphed together with $f(x)$.

```
> plot(y, 1.1..2.2);
```

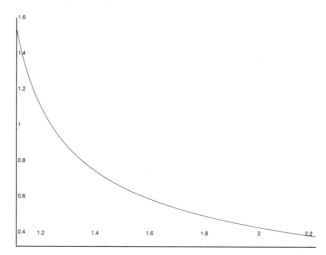

```
> slope:=(subs(x=2.2,y)-subs(x=1.1,y))
                /(2.2-1.1);
            slope := -1.055833850
> c1:=fsolve(diff(y,x)-slope,x,1.1..2.2);
> s:=slope*x+(subs(x=c1,y)-slope*c1);
      s := - 1.055833850 x + 2.231843849
> plot({y,s},1.1..2.2);
```

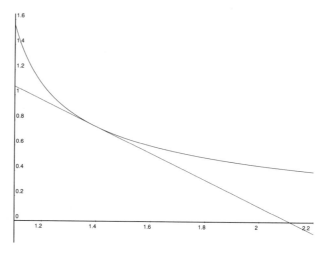

Part 2(c): As suggested in 2(c), the slope of the secant line is calculated followed by an evaluation of the error, $e(x)$. As a check, $x = 1.1$ and 2.2 are substituted in $e(x)$ (these produce 0 and 10^{-9}, respectively). Since the error is not zero for all x in this interval, there must be at least one point where it is an extremum (maximum or minimum). We find all such points by differentiating the error with respect to x and, having plotted this error function and its derivative to get an insight into their behavior, equate the derivative to zero to find the extrema. The graph of the error shows a minimum, with the derivative crossing the x-axis near $x = 1.4$. A better estimate of the location of this minimum is obtained by solving the derivative for x (this is given by c below).

```
> secant:=(subs(x=2.2,y)-subs(x=1.1,y))
                         /(2.2-1.1);
> e:=y-(subs(x=1.1,y)+secant*(x-1.1));
> subs(x=1.1,e);
> subs(x=2.2,e);
> de:=diff(e,x);
                2 x + 2
de:=-1/2 ------------------- + 1.055833850
              2         3/2
          (x  + 2 x - 3)
> c:=fsolve(de=0,x,1.1..2.2);
            c := 1.39276397895
> plot({e,de},1.1..2.2);}
```

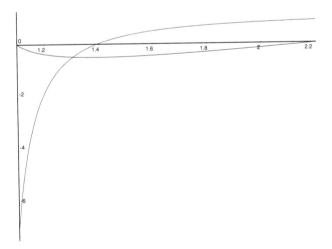

Part 3: We define a parabola in general form and find the value of c such that $f'(c) = (f(b) - f(a))/(b - a)$. After simplifying c, we see that $c = (a + b)/2$ which clearly satisfies $a < c < b$.

```
> y:=a0+a1*x+a2*x^2;
                           2
         y := a0 + a1 x + a2 x
> c:=solve(diff(y,x)=(subs(x=b,y)
       -subs(x=a,y))/(b-a),x);
> c:=simplify(c);
         c := 1/2 a + 1/2 b
```

4   Conclusions

Our principal thesis is that a CAS can provide a comfortable environment in which students can learn mathematics. We are encouraged to continue employing CASs in teaching mathematics by three facts. First, we have seen an improvement in all the traditional indicators used in course evaluations: see Muller [6]. Second, CASs are surely here to stay; and third, the mandate of a university includes the need to expose students to the cutting edge of activity in each discipline: CASs are a new cutting edge in the teaching and learning of mathematics. Students cannot assess the value of CASs in their own education unless they have personally experienced them, so we intend to ensure that our students obtain some familiarity with CASs. In turn, our students, in attempting to follow our directions for the use of these systems, have occasionally (either through ingenuity, or by making a mistake that we had not foreseen) opened new avenues of application. We know much less than we would like to know about the ways in which students actually learn mathematics, and we therefore feel it is very important not to close any likely avenues of learning for them. Students and teachers are all in unfamiliar territory, and interactions between motivated students and concerned teachers will provide new excitement and learning experiences. We believe that **we no longer have the right** to withhold the CAS environment from our students.

References

1. Auer, J. W. (1991). *Linear Algebra with Applications*, Prentice-Hall Canada Inc., Scarborough, Ontario, Canada.

2. Auer, J. W. (1991) *Maple Solutions Manual for Linear Algebra with Applications*, Prentice-Hall Canada Inc., Scarborough, Ontario, Canada.

3. Hillel, J. (1991). *Computer Algebra Systems as Learning Tools*, ZDM, vol 23, No. 5, 184–191.

4. Hodgson, B. R. and Muller, E. R. (To appear). "The Impact of Symbolic Mathematical Systems on Mathematics Education," in Cornu, B. and Ralston, A. (eds.), *The Influence of Computers and Informatics on Mathematics and Its Teaching*, UNESCO.

5. Karian, Z. (1990). "Symbolic Computation: A Revolutionary Force," *UME Trends*, August.

6. Muller, E. R. (1991). "*Maple* Laboratory in a Service Calculus Course," in Leinbach, C. *et al.* (eds.) *The Laboratory Approach to Teaching Calculus*, MAA Notes Number 20, Mathematical Association of America, 111–117.

7. Page, W. (1990). "Computer Algebra Systems, Issues and Inquiries," *Computers Math. Applic.*, Vol 19, No. 6, 51–69.

8. Symbolic Computation Group, University of Waterloo. (1989). *Maple Symbolic Computation for the Macintosh*, Brooks/Cole Publishing Co., Pacific Grove, California.

9. Tall, D. (1991). "Recent Developments in the Use of Computers to Visualize and Symbolize Calculus Concepts," in Leinbach, C. *et al.* (eds.) *The Laboratory Approach to Teaching Calculus*, MAA Notes Number 20, Mathematical Association of America, 15–25.

Using Graphing Calculators to Enhance the Teaching and Learning of Linear Algebra

Don LaTorre
Clemson University

1 Introduction

An introductory course in linear algebra is standard fare in the undergraduate mathematical education of students at virtually every college and university. Directed primarily towards those in technical fields, such courses range from the lightweight to the mathematically rigorous, with varying degrees of emphasis on theory and applications. Not only does the material have widespread applicability, but it has served us well for many years as an introduction to the theoretical and axiomatic approach characteristic of advanced study in the mathematical sciences.

Until recently, however, beginning courses were constrained to stick largely to the theory of linear algebra, provided little experience with matrices that were not contrived, and generally ignored (sometimes to the point of conveying misconceptions) even the most basic numerical considerations that are almost always present in real applications. But today's computer algebra systems (CAS) are changing all that. With easy-to-use technology now increasingly available to students, there is a growing awareness that its careful and judicious use can do much to enhance the teaching and learning of linear algebra at all levels.

Within the last two years, high-level graphing programmable calculators with symbol manipulation capabilities (the Hewlett-Packard 28S and 48S) have exploded into undergraduate mathematics. For the first time, our students have real graphical, numerical and symbolic computing power, more power than the campus mainframes of 25 years ago, in the palms of their hands. Though most of the early interest in these devices has been directed towards their use in calculus, the calculators are also proving to be an attractive

choice of CAS to use in introductory linear algebra.

This article is a reflection of my experiences teaching an introductory, sophomore-level course in matrix-oriented linear algebra since August 1989, a course in which all students have their own HP-28S or HP-48S calculator. My intentions here are:

1. Establish that the Hewlett-Packard units are a good, if not ideal, choice of CAS to use in such courses,

2. Illustrate some of their more appropriate pedagogical uses.

2 The Role of Calculators

The benefits of using CAS in introductory linear algebra are readily identified:

1. Remove the computational burden often associated with hand performance of matrix algorithms, thus allowing beginning students to focus more clearly on the underlying concepts and theory;

2. Encourage and enable students to engage in some exploratory/discovery work on their own;

3. Consider more interesting and realistic applications;

4. Begin a discussion of some of the computational aspects of linear algebra;

5. Demonstrate some of the advantages and power of technology in mathematics.

Although graphing and programmable calculators are technically inferior to many microcomputer-based computer algebra systems in terms of speed, image resolution, display size, and software flexibility, it is not at all clear that the level of technological sophistication found in the latter is either desirable or necessary to achieve the aforementioned benefits in an introductory course. To the extent that the HP-28S and the HP-48S can help achieve these benefits in a fairly substantial way, they are an entirely appropriate choice of CAS and certainly are not inferior to computer-based systems in a pedagogical sense. And it is the pedagogical and learning issues with which we are primarily concerned.

When students are equipped with their own calculators, they use them almost daily for homework and classwork and there is immediate feedback as well as a strong element of participation and interaction. My lecturing has given way to a more informal discussion interspersed with appropriate explanations and examples, which students seem to find more interesting.

The presence of calculators introduces a new dynamics into the learning process: students generally seem to become more actively involved in their thinking about the material. My students see them as especially applicable to their needs, for they work equally well in hallways, in the library, at park benches and lab benches, and are a constant companion in their backpacks. They need not confine their explorations to central facilities nor individually spend a considerable amount of money on a computer. There is a genuine aura of excitement surrounding their use, which can only be interpreted in a positive sense.

Basic keystroke commands are used whenever possible (e.g., matrix addition and multiplication, scalar multiplication, transposes, dot products and norms). I provide the class with a variety of simple programs, each of which is short and addresses a specific aspect of the course. To avoid overwhelming beginners with technical detail, I avoid a line-by-line explanation of the coding. And, though these programs may appear cryptic as text, they are far less cryptic as keystrokes. They simply record a direct, step-by-step solution to a specific problem.

I also make no attempt to incorporate professional-level code and am careful not to use programs that present final results at the expense of involving students with the underlying mathematical processes. At this level, I am primarily interested in increasing students' interest, involvement, comprehension and retention of the course material. Since beginning students typically have diverse backgrounds and experiences, care must be taken not to let the technology overpower the course and steer it in inappropriate directions. However, professional-level code for linear algebra will soon be available for the HP-48SX on the plug-in 32K or 128K RAM cards that it accepts. Its availability will most certainly present new and challenging opportunities to use these calculators not only in introductory courses, but also at more advanced levels.

3   Themes and Opportunities

In addition to routine matrix arithmetic, determinants and inverses, there are several major themes in introductory linear algebra which are well-suited to serious CAS enhancement:

1. Gaussian elimination and LU-factorizations

2. Vector space theory associated with matrices

3. Geometrical notions and orthogonality

4. Eigenvalues and eigenvectors.

Consequently, my calculator-enhanced course is structured around these themes, and opportunities for enrichment abound. Iterative methods, for example, extend the traditional material and point the way toward the powerful numerical approaches used to solve large linear systems and to obtain eigenvalues and eigenvectors. However, I have resisted the temptation to restructure the course into one emphasizing the computational aspects of numerical linear algebra.

The following sections present some specific examples of how the HP-28S and HP-48S units may be used in the context of these themes to achieve the pedagogical benefits cited earlier. For a more detailed discussion, see [1].

4   Determinants and Inverses

One of the common misconceptions that students have traditionally carried away from introductory linear algebra concerns the role of determinants, especially with regard to matrix inverses. Equipped with the elementary result that matrix A is invertible if and only if $\det A \neq 0$, they are likely to blindly apply the result as a computational test for invertibility. And with determinants available upon a single command, why shouldn't they?

As a simple experimental exercise which is done in class by everyone, students use their HPs to calculate the determinants and inverses of both

$$A = \begin{bmatrix} 1 & 1 & 1 \\ 2 & 4 & 6 \\ 2 & 4 & 6 \end{bmatrix}$$

and

$$B = \begin{bmatrix} 1 & 1 & 1 \\ 3 & 6 & 4 \\ 3 & 6 & 4 \end{bmatrix}$$

As expected, the calculators show $\det A = 0$ (because A has two identical rows), and the message "INV Error: Infinite Result" appears when students, improperly, ask for A^{-1}. But $\det B$ appears as 3×10^{-11}, so when (properly?) asked for B^{-1}, they see

$$B^{-1} = \begin{bmatrix} 2 & 66666666666.6 & -666666666667 \\ -1 & 33333333333.3 & -333333333333 \\ 0 & -99999999999.9 & 100000000000 \end{bmatrix}.$$

Startled, they quickly check that

$$B^{-1}B = \begin{bmatrix} .8 & -.4 & .4 \\ -.1 & .8 & .2 \\ .3 & .6 & .4 \end{bmatrix}$$

and

$$BB^{-1} = \begin{bmatrix} 1 & 0 & 0 \\ 0 & 0 & 0 \\ 0 & 0 & 1 \end{bmatrix},$$

which only makes things worse! Matrix B, like A, has two identical rows. Thus $\det B = 0$ and B has no inverse. So what's going on here? There is surely little in the appearance of B vs. A to suggest these results. The point here is clear: indiscriminate use of the calculators to calculate determinants and matrix inverses may produce incorrect results. In the case at hand, a little forethought would certainly have been in order.

But what if the matrices are not so obviously singular? And why have the calculators failed to return results for B like those for A? The answer to the latter lies both with the built-in routine for finding determinants and inverses (the Crout decomposition) and the floating point environment that the calculators use. Without considering the Crout algorithm in detail, the above determinants come down to calculating

$$\det A = 2[6 - (2 \times 3)],$$

and

$$\det B = 3 \left[4 - \left(3 \times \frac{4}{3} \right) \right].$$

In exact arithmetic, both are 0. The HP's floating point calculation of $\det A$ will be 0, but for $\det B$ we have

$$\begin{aligned} \det B &= 3[4 - (3 \times 1.33333333333)] \\ &= 3[4 - 3.99999999999] \\ &= 3[1 \times 10^{-11}] \\ &= 3 \times 10^{-11}. \end{aligned}$$

And, given that a nonzero value for $\det B$ was returned, the routine went on to calculate an inverse.

The message in this exercise is obvious. Calculating a determinant as a test for matrix invertibility is computationally impractical in a floating point environment. And the size of the determinant also has no bearing on the invertibility, for while multiplying an $n \times n$ matrix A by a nonzero number k does not change the invertibility, the determinant of the resulting matrix is $k^n(\det A)$. In fact, there is little, if any, need to calculate determinants of matrices other than the low order, integer-valued matrices used in elementary courses to reinforce the learning of the basic theory. The reality is that the inclusion of determinants in linear algebra courses is to some extent a carry-over from late nineteenth-century algebra, and aside from their important role in helping to understand certain other concepts, especially those associated with parameter-dependent matrices like $(A - \lambda I)$, they have little practical use—certainly no use in computational mathematics.

5 Systems of Linear Equations

The most popular methods for dealing with linear systems in introductory linear algebra courses are the elimination methods, consisting of several variants of Gaussian elimination with back substitution. Many beginning courses blur the distinction between these variants in the interest of expediency. But with an eye toward subsequent study in linear analysis or numerical methods and the use of professional elimination codes, it is important that students carefully distinguish between the traditional Gaussian elimination algorithm, the back substitution process, partial pivoting and Gauss-Jordan reduction. Likewise, it is important to understand Gaussian elimination for square matrices as a factoring process which factors a matrix A into triangular factors, $A = LU$. These several aspects of linear systems are all nicely suited to calculator enhancement and are good examples of where the HPs can be used to substantially remove the computational burden associated with paper-and-pencil methods.

In its traditional from, the Gaussian elimination algorithm for solving a linear system $Ax = b$ adds suitable multiples of one equation to the others with the goal of obtaining an equivalent system $Ux = b'$, where the augmented matrix $[U|b']$ is in row-echelon form. It may be necessary to interchange equations at various times for the elimination process to continue. Back substitution then solves $Ux = b'$ systematically. Carried out without reference to the unknowns by working with the augmented matrices $[A|b]$ and $[U|b']$, computationally the only source of error is round-off, induced by the computational device itself.

Once students have acquired a basic understanding of Gaussian elimination and have worked several examples by hand, the HPs can be used to efficiently perform the row operations which transform $[A|b]$ into $[U|b']$. They are given program ELIM, which pivots on a specified entry, the pivot, to produce 0's below that entry. Written to handle both real and complex matrices, the program is used, more generally, to convert a matrix to row-echelon form. In the event the intended pivot is 0, the program stops and prints the error message "PIVOT ENTRY IS 0". To be genuinely useful, program ELIM must be used with a row-swapping routine, RO.KL, and a simple routine BACK for carrying out the back substitution process.

All these programs are interactive and require input and control at key steps. ELIM requires the students to decide when and where to pivot, and to use RO.KL they must specify which row interchanges are needed. BACK performs the back substitution one step at a time, each step requiring the press of a single key. By carrying all the computational burden, these routines let students focus more clearly on the elimination and back substitution processes. And they have complete control from start to finish.

With row interchanges we are simply avoiding 0 pivots. But, for the practical real-world solution of large-scale linear systems, it is just as important to avoid using extremely small pivots, because division by small numbers in floating point arithmetic may ultimately induce considerable error. The common pivoting strategy known as partial pivoting chooses as the pivot any element in the pivot column whose absolute value is maximum. The need for this strategy is awkward to illustrate on the HPs because of

their use of 12 digit mantissas. Nevertheless, I require that my students adopt the strategy often to reinforce their understanding of it. This is easy to do with the calculators but almost impossible without them.

Students quickly show mastery of the above, but are left with no real power tool to completely solve small linear systems. To remedy this situation and at the same time provide a tool that is extremely useful for both classwork and homework dealing not only with linear systems *per se*, but also other concepts associated with such systems (e.g., linear independence, change-of-basis and eigenvectors), I permit my students to use an interactive routine GJ.PV which implements the popular variant of Gaussian elimination known as Gauss-Jordan reduction. Though Gaussian elimination with back substitution is more efficient than Gauss-Jordan reduction for dealing with linear systems in general, and is certainly the preferred method in professional computer libraries, most students prefer to use Gauss-Jordan reduction for the small-scale problems encountered in the course.

Gauss-Jordan reduction differs from Gaussian elimination in two ways:

1. All the pivots are converted to 1;

2. The basic pivot process is used to produce 0's both below and above the pivot element.

When applied to a nonzero matrix A, Gauss-Jordan reduction yields what is popularly called the reduced row echelon form (RREF) of A:

1. Any zero rows lie at the bottom;

2. The first nonzero entry in any nonzero row (the pivot) is a 1, and lies to the right of the pivot entry in any preceding row;

3. The pivot entry is the only nonzero entry in its column.

The reduced row echelon form of A is important because it represents the ultimate we can get from A by applying elementary row operations. As such, it is uniquely associated with A.

I have found the Gauss-Jordan reduction program (program GJ.PV) to be very effective in helping students focus upon the role of free variables vs. pivot variables. Consequently, it has proven to be a valuable pedagogical tool which they can readily use in working with linear systems to help understand some of the more abstract concepts encountered in the course: linear combinations and spanning sets, dependence and independence, bases and dimension, change of basis and eigenvectors. Though these concepts initially may appear to be somewhat removed from linear systems, exactly the opposite is true: it was from a study of linear systems and their associated matrices that these concepts emerged.

In addition to recognizing Gaussian elimination as an orderly process for converting a square matrix to upper triangular form, it is important that students also understand it as a factorization process, the so-called LU-factorization. This viewpoint is not only interesting from a purely algebraic standpoint, it also lies at the heart of many computer codes. In its simplest form, when the matrix A in a linear system $Ax = b$ can be brought to upper triangular form U by Gaussian elimination without row interchanges, then $A = LU$ where L is lower triangular with 1's along its diagonal and the entries below the diagonal are the negatives of the multipliers employed in the elimination process. When row interchanges are needed to avoid 0 pivots, then $A = LU$ is no longer valid; it is replaced by a factorization of the form $PA = LU$ where P is a permutation matrix that accounts for the row interchanges. Since I am primarily interested in the pedagogical aspects of LU-factorizations I have chosen to use a calculator program, LU (given at the end of this article along with a worked-out example), which the students must control at each step just as in the case of hand calculations. With LU-factorizations readily available, my students have no trouble in applying them to solve linear systems $Ax = b$ with multiple right-hand sides. Equipped with all the basic ingredients from $PA = LU$, they use program FWD to do the forward substitution in $Ly = Pb$ needed to get vector y, then apply program BACK to do the back substitution in $Ux = y$ needed to produce the solution x.

6 Orthogonality Concepts

Much of the geometry of $\Re^2$ and $\Re^3$ can be extended to $\Re^n$ by means of the standard inner product (dot product) in $\Re^n$: length, distance, projections, and orthogonality. Ultimately, these notions all come to focus on the really important one, orthogonality, and my introductory course devotes serious attention to it. Orthogonal subspaces, orthonormal bases, and orthogonal projections are difficult concepts for beginning students. But they are needed because, when

cast in the context of matrices, they have important numerical applications.

The symbolic manipulation and numerical computation capabilities of the HP-28S and HP-48SX are genuinely helpful to students at this point, especially as they work to understand orthogonal subspaces, the Gram-Schmidt orthonormalization process, its interpretation as a QR-factorization, and the application of all these ideas to linear least squares problems. QR-factorizations, though often omitted from introductory courses, are crucial because, like LU, they are central to many professional codes for linear algebra.

Here is a typical calculator-based exercise designed to reinforce students' understanding of the fundamental result that, given a matrix A, the nullspace NS(A) is the orthogonal complement of the rowspace RS(A) and the nullspace of A^T, NS(A^T), is the orthogonal complement of the column space CS(A). Note that the exercise addresses a number of concepts: subspaces, bases, linear systems, the rank-nullity theorem, and orthogonality. Without serious computational assistance, provided here by the basic Gauss-Jordan reduction program GJ.PV, the exercise would be too laborious to complete by hand.

Exercise. For the matrix A below

$$A = \begin{bmatrix} 1 & 2 & 2 & -1 & 2 & 2 \\ 0 & 0 & 1 & -1 & 1 & 0 \\ 1 & 2 & 2 & 0 & 1 & 3 \\ 2 & 4 & 3 & -2 & 4 & 2 \\ 1 & 2 & 1 & -2 & 3 & 2 \end{bmatrix}$$

1. Find a basis for NS(A) and verify that its vectors are orthogonal to those in a basis for RS(A);

2. Find a basis for NS(A^T) and verify that its vectors are orthogonal to those in a basis for CS(A).

After orthogonal projections have established the desirability of an orthonormal basis for a subspace W of $\Re^n$, we study the traditional Gram-Schmidt process for constructing such a basis $q_1, q_2, \ldots, q_k$ from a given basis $x_1, x_2, \ldots, x_k$ for W. Recall how it works.

Let q_1 be the normalized version of x_1,

$$q_1 = \frac{x_1}{\|x_1\|}.$$

Then, inductively, having constructed orthonormal vectors $q_1, \ldots, q_j$, we construct q_{j+1} as follows:

$$q_{j+1} = \ x_{j+1} - \text{(the sum of the projections of } x_{j+1} \text{ onto } q_1, q_2, \ldots, q_j,), \text{ normalized.}$$

Thus, before normalization, $q_{j+1} = x_{j+1} -$ (the projection of x_{j+1} onto the subspace spanned by $q_1, \ldots, q_j$). Let's look at several steps:

1: $q_1 = x_1$, normalized

2: $q_2 = x_2 - \underbrace{(x_2 \bullet q_1)q_1}_{\text{proj. of } x_2 \text{ onto } q_1}$, normalized

3: $q_3 = x_3 - \underbrace{(x_3 \bullet q_1)q_1 - (x_3 \bullet q_2)q_2}_{\text{proj. of } x_3 \text{ onto } q_1 \text{ and } q_2}$, normalized

$\vdots \quad \vdots$

Relative to using the HPs, my position is that it is not pedagogically sound for beginning students to use a program that does it all and hides the underlying algorithm. I prefer instead that they prepare and evaluate a simple program, using stored vectors as inputs and basic keystroke commands to carry out each step of the construction. There are two reasons for this: first, in the course of entering the programs they demonstrate and reinforce their understanding of the algorithm; and second, they can easily review their work before execution and thereby avoid procedural errors. They begin with the original basis vectors stored as variables X1, X2, ..., XK, and use a one-line program PROJ which calculates the projection of one vector onto another.

1: ≪ X1 X1 ABS / ≫ | EVAL | | ′ | Q1 | STO |
   (calculates q_1 and stores it as Q1)

2: ≪ X2 X2 Q1 PROJ - DUP ABS / ≫
   | EVAL | | ′ | Q2 | STO |
   (calculates q_2 and stores it as Q2)

3: ≪ X3 X3 Q1 PROJ - X3 X2 PROJ - DUP
   ABS / ≫ | EVAL | | ′ | Q3 | STO |
   (calculates q_3 and stores it as Q3)

$\vdots \quad \vdots \qquad \vdots$

Here, the calculator is simply recording the symbolic manipulations of the algorithm and then performing the numerical calculations called for. The students themselves specify the procedures and maintain complete control over the process, step-by-step.

Not only do I require my students to understand that Gaussian elimination applied to a matrix A amounts to an LU-factorization $A = LU$, I also require them to understand that the Gram-Schmidt process applied to a matrix A having independent columns amounts to a QR-factorization $A = QR$

where Q has orthonormal columns and R is upper triangular and invertible.

We establish the $A = QR$ result carefully (the columns of Q are the orthonormal q_i's produced by the Gram-Schmidt process and the (i, j)-entry of R is given by $r_{ij} = q_i \bullet x_j$). But having already effectively used the calculators to construct the column vectors of Q, it is now just a simple matter to continue and assemble these vectors into the Q matrix, build matrix R with a few dot products, and then check (after cleaning up the inevitable round-off error) that $A = QR$.

My point here regarding calculator use is simply this: before the calculators, my students were almost never able to get to the QR aspect because there were so few Gram-Schmidt problems that they could work by hand without getting bogged-down in arithmetical computation and error. Indeed, all except the very best students became so overwhelmed by hand calculations that they never really mastered the orthonormalization process. Freed now from paper- and-pencil methods, they not only master orthonormalization and go on to obtain QR-factorizations, but actually apply them to linear least squares problems.

In a typical linear least squares problem we are concerned with fitting a polynomial

$$P(x) = c_0 + c_1 t + \cdots + c_m t^m$$

of degree $\leq m$ to n data points

$$(x_1, y_1), (x_2, y_2), \ldots, (x_n, y_n)$$

where all the x_j's are distinct and $n \geq m + 1$. Thus we are led to consider a linear system $Ac = y$ where

$$A = \begin{bmatrix} 1 & x_1 & x_1^2 & \cdots & x_1^m \\ 1 & x_2 & x_2^2 & \cdots & x_2^m \\ & & \vdots & & \\ 1 & x_n & x_n^2 & \cdots & x_n^m \end{bmatrix},$$

$$c = \begin{bmatrix} c_0 \\ c_1 \\ \vdots \\ c_m \end{bmatrix} \text{ and } y = \begin{bmatrix} y_1 \\ y_2 \\ \vdots \\ y_n \end{bmatrix}.$$

Since $n \geq m + 1$, there are at least as many equations as unknowns and the system will, in general, be overdetermined. We naturally seek a least squares solution. Matrix A has independent columns, so there is a unique least squares solution, given as the unique solution to the normal equations

$$(A^T A)c = A^T y.$$

The coefficient matrix $A^T A$ is often ill-conditioned so we do not want to calculate $(A^T A)^{-1}$. Applying $A = QR$, $Q^T Q = I$ and the fact that R is invertible, the normal equations reduce to $Rc = Q^T y$ which is readily solved by back substitution.

Students engage calculator-based activities like this:

1. Generate a random 5×4 matrix A over Z_{10} and obtain a QR-factorization of A. Check your results.

2. Now generate a random 5×1 vector b over Z_{10} and use your QR-factorization of A to construct a least squares solution to $Ax = b$.

Although we use the traditional Gram-Schmidt process for our QR-factorizations, I caution students that this process is numerically unstable in floating point arithmetic. That is, round-off errors usually conspire to produce vectors that are not, numerically, orthogonal. Although there is a variation of the algorithm that is more stable, I avoid it in my course because it is not geometrically obvious. I tell them that, in practice, other methods are used: Householder reflections or Givens rotations. These are orthogonal matrices which can be used very effectively to obtain QR-factorizations.

7 Eigenvalues and Eigenvectors

Eigenvalues and eigenvectors are the last of the themes around which my introductory course is constructed.

Traditionally, students are required to work through a variety of elementary problems as they begin to meet eigenvalue-eigenvector notions to help reinforce their understanding of the ideas. But, more often than not, the hand calculations that are required soon overwhelm all but the most able of them, and many never move much beyond the basics. Mired in calculations, they do not get a real chance to seriously consider what is often the last major objective of the course: the orthogonal diagonalization of a real symmetric matrix.

The HPs help by substantially removing the computational burden associated with hand calculation of characteristic polynomials, eigenvalues and associated eigenvectors, and the construction of an orthogonal diagonalizing matrix Q. My students use a short program, CHAR, that calculates the coefficients of

$$\det(\lambda I - A) = \lambda^n + c_{n-1}\lambda^{n-1} + \cdots + c_1\lambda + c_0,$$

which is the characteristic polynomial of A, or $(-1)^n$ times the characteristic polynomial of A, depending upon your point of view. The program implements the SOURIAU-FRAME method, which uses traces of the first n powers of A. Since the characteristic polynomial of a matrix with integer entries has integer coefficients, CHAR returns these coefficients without error for the matrices we normally use in the course: random matrices over Z_{10} of modest size, say, $n \le 6$.

They first use this program with another one, P.OFA, to do an experiment that leads them to discover the Cayley-Hamilton theorem: every square matrix satisfies its characteristic polynomial. CHAR returns the characteristic polynomial $p(\lambda)$ of A and P.OFA evaluates $p(A)$. Actually, this experiment is one of several such discovery exercises that we use in connection with characteristic polynomials. Others are directed toward discovering how the trace of A and $\det A$ appear in the characteristic polynomial and how the trace of A and $\det A$ are related to the eigenvalues of A. See the next section for details.

To avoid the highly contrived matrices that tend to dominate introductory-level texts, I encourage students to use a polynomial root-finder program to obtain the eigenvalues of matrices whose entries are integers as the roots of the characteristic polynomial. Bill Wickes of Hewlett-Packard has included such a routine, PROOT, in [2]. PROOT implements a closed-form technique for finding all roots of polynomials of degree ≤ 4. Though there is no closed-form formula for the roots of a general polynomial of degree 5, the calculator's solver can be used to find a real root x_0, a polynomial divide routine can be used to factor out $x - x_0$, and PROOT can then obtain the other four roots. Because of Perron's result, this process may be extended to positive matrices of higher order, though it can become unwieldy for $n > 6$. Indeed, when my classes were HP-28S based, the students used this PROOT program. However, since the beginning of the 1990 Fall term the classes have been HP-48S based and I have obtained permission to use a more powerful and robust polynomial root-finder program from Hewlett-Packard, written in machine language and not yet available in the public domain, which uses advanced numerical methods to produce all roots of an arbitrary polynomial.

Although low order matrices having integer entries are not typical of the matrices encountered in scientific and engineering applications, they serve us well in the learning process. And, I am quick to remind my students that the above procedures for finding eigenvalues are not appropriate for matrices whose entries are not integers, because it is not always possible to determine the coefficients of the characteristic polynomial with enough accuracy to be sure of the roots. The numerical calculation of eigenvalues is a much more complicated problem than, say, the numerical solution of linear systems, and any discussion of the techniques employed is well beyond an introductory course. But the HPs have, however, enabled me to touch upon some of the key ideas: QR-factorizations and Householder matrices. I refer my better students to Strang's excellent expository account (Section 7.3 in [3]) and to more advanced courses in numerical methods.

In spite of this disclaimer, students seem willing to settle for the eigenvalues of small, integer-valued matrices and are interested in continuing on to a study of the principal axis theorem: each real symmetric matrix A can be converted to a diagonal matrix D by means of an othogonal transformation $Q : Q^T A Q = D$. Our previous discussions on using the HPs to solve linear systems and to build orthonormal bases all apply to the construction of Q.

8 Exploration and Discovery

One of the objectives of using CAS in an introductory course in linear algebra is to encourage students to become active participants in their learning of the material. Typically, they are majoring in a variety of technical fields, most often in computer science, engineering, mathematical sciences, or secondary mathematics education. Mathematically unsophisticated, many are still enrolled in the basic calculus and differential equations course sequence. With the exception of the few who later enroll in an upper-division course in numerical methods or continue on to graduate study in mathematics, an introductory course will probably be their only formal and systematic encounter with linear algebra.

Activities that lead students to explore, conjecture, and discover on their own can be a real catalyst towards learning. In addition to offering the intrinsic appeal of discovery, such experiments tend to foster independent thinking by forcing students to become more than passive recipients of someone else's knowledge. Listed below is a sampling of the exploratory

activities that my students complete. None is complicated, but each one uses the calculators to help students to discover something on their own, as opposed to my telling them or having them read about it from their textbook. They serve as the bases for subsequent classroom discussions.

Markov Matrices. For this exercise, set your calculator to 3 FIX mode. Let

$$A = \begin{bmatrix} .3 & .3 & .3 & .2 \\ .4 & .3 & .2 & 0 \\ .1 & .2 & .2 & .3 \\ .2 & .2 & .3 & .5 \end{bmatrix},$$

and

$$x = [.1 \ .2 \ .3 \ .4]^T.$$

1. Examine the sequence $A, A^2, A^3, \ldots$ to find $\lim_{n \to \infty} \{A^n\}$.

2. Examine the sequence $Ax, A^2 x, A^3 x, \ldots$ to find $\lim_{n \to \infty} \{A^n x\}$.

3. What is the connection between the two limits in Parts 1 and 2?

4. Repeat Parts 2 and 3 using any vector $x = [a \, b \, c \, d]^T$ of your choice where $a + b + c + d = 1$.

5. Prepare a written report summarizing your investigations.

Outer Products. Generate and store a random 4×3 matrix A and a random 3×5 matrix B, both over Z_{10}.

1. Separate A into its 4×1 column matrices and store them in their natural order as A_1, A_2 and A_3.

2. Separate B into its 1×5 row matrices and store them in their natural order as B_1, B_2 and B_3.

3. Calculate the matrix $A_1 B_1 + A_2 B_2 + A_3 B_3$ and compare with the matrix product AB.

4. Repeat Parts 1 through 3 using the columns and rows of a random 3×4 matrix A and a random 4×5 matrix B over Z_{10}.

5. Summarize your findings and formulate a conjecture based upon them, being sure to write in complete English sentences. Be prepared to hand-in your write-up and to discuss it in class. (Note: each of the products $A_i B_i$ is called an outer product.)

LU-Factorizations. For $n = 4$, 5 and 6

1. Generate a random (invertible) $n \times n$ tridiagonal matrix A over Z_{10}, and find A^{-1}.

2. Obtain an LU-factorization of A and note any specialized structure of L and U.

3. Prepare a written report in which you summarize the results of your investigations in Parts 1 and 2.

Least Squares.

1. Use FEVAL to fill in the following table of values for $f(x) = (x + 2)^2 e^{-x}$ (round to 3 decimal places).

x	-2.2	-1.0	0.5	1.5	3.0
$f(x)$					

2. Use DPLOT to plot the 5 data points.

3. Find the least squares cubic polynomial fit, $P_3(x)$, to this data; overlay your data plot with the graph of $P_3(x)$.

4. Find the interpolating polynomial $P_4(x)$ for this data; overlay your data plot with the graph of $P_4(x)$.

5. Be prepared to discuss your work with your instructor.

Orthogonality.

1. Generate a random 4×3 matrix over Z_{10} whose columns will be called x_1, x_2, and x_3.

2. Construct an orthonormal basis $\{q_1, q_2\}$ for $W = \mathrm{Span}([x_1 \; x_2])$.

3. Find the projection vector $P_W x_3$ of x_3 onto W.

4. Verify that $x_3 - P_W x_3$ is orthogonal to W by checking that it is orthogonal to both x_1 and x_2.

5. Repeat Parts 1 through 4 with a random 5×4 matrix over Z_{10}; let $W = \mathrm{Span}([x_1 \; x_2 \; x_4])$.

Traces.

1. Generate a random 3×4 matrix A and a random 4×3 matrix B, both over Z_{10}.

2. Compare trace(AB) and trace(BA); what do you observe?

3. Repeat Parts 1 and 2 for random 3×5 and 5×3 matrices over Z_{10}.

4. Formulate a conjecture on the basis of your observations. Prove your conjecture.

Cayley-Hamilton Theorem.

1. Generate a random 3×3 matrix A over Z_{10} and put two copies of A on the stack.

2. Find the characteristic polynomial $p(x)$ of A and evaluate $p(A)$.

3. Repeat Parts 1 and 2 using random 4×4 and 5×5 matrices over Z_{10}.

4. Repeat Parts 1 through 3 using random 3×3, 4×4, and 5×5 complex matrices over Z_{10}.

5. Formulate a conjecture based upon your observations. Discuss your conjecture with your instructor.

Traces and Eigenvalues. For each of the following matrices A,

$$\begin{bmatrix} 1 & 1 & 2 \\ 2 & 1 & 1 \\ 2 & 1 & 2 \end{bmatrix}, \quad \begin{bmatrix} 5 & 1 & 4 \\ -6 & 2 & 6 \\ 0 & 0 & -1 \end{bmatrix},$$

$$\begin{bmatrix} 7 & 2 & 4 & 6 \\ -6 & -1 & -4 & -4 \\ 4 & 4 & 5 & -2 \\ -16 & -12 & -14 & -3 \end{bmatrix},$$

and

$$\begin{bmatrix} -14 & -16 & -26 & -9 \\ 16 & 19 & 28 & 12 \\ -7 & -8 & -11 & -7 \\ 13 & 14 & 24 & 14 \end{bmatrix}.$$

1. Calculate the trace of A (tr A) and det A; record your results.

2. Compare tr A and det A with $\det(\lambda I - A)$; what do you observe? Express your observation as a conjecture.

3. Compare tr A with the sum, $\sum_i \lambda_i$, of the eigenvalues of A; what do you observe? Express your observation as a conjecture.

4. Compare $\prod_i \lambda_i$, the product of the eigenvalues of A, with $\det A$; what do you observe? Express your result as a conjecture.

5. Discuss your conjectures with your instructor.

Iterative Solutions. Consider the tridiagonal system

$$
\begin{array}{rrrrrrr}
x_1 & -x_2 & & & & & = -4 \\
-x_1 & +3x_2 & -x_3 & & & & = 11 \\
& -x_2 & -3x_3 & -x_4 & & & = -1 \\
& & -x_3 & +3x_4 & -x_5 & & = 3 \\
& & & -x_4 & -3x_5 & -x_6 & = -2 \\
& & & & -x_5 & +x_6 & = -3
\end{array}
$$

1. Apply all our tests for convergence. What can you conclude?

2. Remember, these tests are only sufficient conditions for convergence. Thus, ignore the test results and try for an iterative solution anyway—accurate to 4 decimal places.

Power Method.

1. Our presentation of the power method required a unique dominant eigenvalue λ_1 (of multiplicity $m \geq 1$). What happens if, say, λ_1 and its negative $-\lambda_1$ dominate? That is, if

$$
|\lambda_1| = |-\lambda_1| > |\lambda_3| > \cdots > |\lambda_n|?
$$

To find out, apply the power method to the following matrix A, using $y_0 = [1\,1\,1\,1]$ and 5 decimal place accuracy. Watch the vector iterations very carefully and count the number of iterations.

$$
A = \begin{bmatrix}
3 & -1 & 0 & 0 \\
-1 & -3 & -1 & 0 \\
0 & -1 & 3 & -1 \\
0 & 0 & -1 & -3
\end{bmatrix}.
$$

2. Has the sequence of vectors converged to a unique approximate eigenvector before reaching the maximum allowable number of iterations? Repeat the iteration as necessary to see again what is happening.

3. Formulate a conjecture based upon your observations for discussion with your classmates and instructor.

9   Student Perceptions

The impact of the calculators has been noticeable and overwhelmingly positive. Students generally say the calculators are user-friendly, perhaps because they are hand-held and menu-driven. When they are used to help obtain a result, students often exhibit a strong sense of "personal ownership" of that result. Indeed, it may well be the highly personalized nature of the HPs which, in addition to their portability and flexibility, makes them so attractive.

The calculator-enhanced course has been taught for five semesters at the time of this writing. Students' reactions have been obtained from personal interviews with the classes by an outside evaluator and from their responses to an extensive written evaluation prepared by the evaluator. Here is a summary, based on 120 student responses to several key questions on that evaluation.

Questions	% who Agree or Strongly Agree
1. The graphics calculator helped me understand the course material.	75
2. The graphics calculator was useful in solving problems.	98
3. The graphics calculator allowed me to do more exploration and investigation.	82
4. The graphics calculator helped me have better intuition about the material.	61

In contrast, 80% of the respondents disagreed or strongly disagreed with the statement "I could have learned more if I had **not** used a graphics calculator."

In the Fall 1990 term, the class size was 44 and 26 students had the HP-48SX while the remaining 18 had the HP-28S. Of this group, 74% expressed strong agreement with the statement " I want to study more courses using a graphics calculator."

10   Conclusion

CAS in the form of high level Hewlett-Packard calculators can do much to enhance both instruction and learning in beginning courses in linear algebra. Not only do the calculators enable us to change what we teach and how we teach it, they also change the tests and the testing environment. No longer must we avoid matrices with complex number entries. And, instead of asking students to construct an orthonormal basis, we can give them an overdetermined linear system $Ax = b$ and ask for a QR-factorization of A which then is to be applied to obtain a least squares solution. Changing testing is mandatory, because if students sense that our testing does not accurately reflect what we do on a daily basis, they will quickly cut back. Changing testing is easy and natural with calculators, but difficult to do with microcomputers.

The mathematical community has not yet reached general agreement as to what should constitute a first linear algebra course, especially the extent to which such a course should incorporate ideas from numerical linear algebra. We still see introductory texts that convey inappropriate and sometimes misleading information about determinants, matrix inverses and Cramer's rule, make LU-decompositions optional, underemphasize orthogonality, and fail to mention QR.

Against this unsettled background, I believe that the introduction of modest technology in the form of high-level calculators can do much to effect some needed change in content, captivate and engage student interest, and enhance both the teaching and learning of elementary linear algebra.

References

1. LaTorre, D. R. (1992). *Calculator Enhancement for Linear Algebra*, Saunders College Publishing, Philadelphia, PA.

2. Wickes, W. C. (1988). *Mathematical Applications*, Hewlett-Packard, Corvallis, OR.

3. Strang, G. (1988). *Linear Algebra and its Applications*, 3rd Ed., Harcourt Brace Jovanovich, San Diego, CA.

Appendix

In this appendix we present some of the details of how the HP calculators can be effectively used by students to obtain LU-factorizations $A = LU$ or, more generally, $PA = LU$.

Calculator program LU, given below, carries the computational burden, and works equally well with real or complex matrices. In addition to performing the basic elimination step, LU stores the negatives of the multipliers below the diagonal in a matrix called "ELL", which initially is the zero matrix. Program

MAKL creates the initial ELL and the initial matrix P. If row interchanges are needed, the proper use of the row swapping program RO.KL must be made with ELL, U and P in order to continue. At the end, the screen display shows U, the lower triangle of L is stored in matrix ELL, and P is stored in user memory.

LU   (Used to get LU-factorizations.)
  Inputs: As a stored variable: a variable 'ELL', obtained from program MAKL (below) and containing a zero matrix.
     level 3: a square matrix A
     level 2: an integer K
     level 1: the integer K
  Effect: Pivots on the (K, K)-entry to return a row-equivalent matrix with 0's below the pivot; also puts the negatives of the multipliers into column K of ELL below the diagonal. Press ⃞ELL⃞ to view ELL. Used iteratively to obtain an LU-factorization.

≪ → A K L ≪ IF 'A(K,L)' EVAL 0 ==
THEN "PIVOT ENTRY IS 0" ELSE A SIZE
1 GET → M ≪ M IDN 'A(1,1)' EVAL TYPE
IF THEN DUP 0 CON R → C END K
M FOR I 'A(I, L)' EVAL {I K} SWAP PUT
NEXT INV {K K} 1 PUT DUP A * SWAP
K 1 + M FOR I DUP {I K} GET NEG 8
RND 'ELL(I, K)' STO NEXT DROP ≫ 8
RND END ≫ ≫

NOTE: For the 28S version, replace both occurrences of 8 RND with 8 FIX RND STD.

MAKL   (Make ELL and P.)
  Input: level 1: a square matrix A
  Effect: Creates a variable ELL containing a zero matrix, and a variable P containing an identity matrix, both the same size as A. Used as the initial start-up to obtain an LU-factorization.

≪ DUP 0 CON 'ELL' STO DUP IDN 'P' STO ≫

Example. Produce an LU-factorization of

$$A = \begin{bmatrix} 2 & 3 & -1 & 2 \\ -4 & -6 & 2 & 1 \\ 2 & 4 & -4 & 1 \\ 4 & 8 & 2 & 7 \end{bmatrix}$$

with partial pivoting.

1. Enter A onto level 1, press $\boxed{\text{MAKL}}$ to create appropriate starting matrices ELL and P, then swap rows 1 and 2 of A with 1, 2 $\boxed{\text{ROKL}}$ to see

$$\begin{bmatrix} [-4 & -6 & 2 & 1] \\ [2 & 3 & -1 & 2] \\ [2 & 4 & -4 & 1] \\ [4 & 8 & 2 & 7]] \end{bmatrix}.$$

Bring P to level 1 with $\boxed{\text{P}}$, make the same row interchange and store the result in P. Press 1, 1 $\boxed{\text{LU}}$ to see

$$\begin{bmatrix} [-4 & -6 & 2 & 1] \\ [0 & 0 & 0 & 2.5] \\ [0 & 1 & -3 & 1.5] \\ [0 & 2 & 4 & 8]] \end{bmatrix}.$$

2. Since the (2, 2)-entry of this last matrix is 0, we must interchange row 2 with some lower row, and partial pivoting says to use row 4. Thus press 2, 4 $\boxed{\text{RO.KL}}$ to effect the interchange, then bring ELL to level 1 with $\boxed{\text{ELL}}$, make the same row interchange and store the result in ELL. Bring P to level 1 with $\boxed{\text{P}}$, make the same row interchange and store the result in P.

3. Now execute 2, 2 $\boxed{\text{LU}}$ to see $U =$

$$\begin{bmatrix} [-4 & -6 & 2 & 1] \\ [0 & 2 & 4 & 8] \\ [0 & 0 & -5 & -2.5] \\ [0 & 0 & 0 & 2.5]] \end{bmatrix}.$$

4. Recall ELL =

$$\begin{bmatrix} [0 & 0 & 0 & 0] \\ [-1 & 0 & 0 & 0] \\ [-.5 & .5 & 0 & 0] \\ [-.5 & 0 & 0 & 0]] \end{bmatrix}.$$

Get L =

$$\begin{bmatrix} [1 & 0 & 0 & 0] \\ [-1 & 1 & 0 & 0] \\ [-.5 & .5 & 1 & 0] \\ [-.5 & 0 & 0 & 1]] \end{bmatrix}.$$

with 4 $\boxed{\text{IDN}}$ $\boxed{+}$, then do $\boxed{\text{SWAP}}$ $\boxed{*}$ to see LU =

$$\begin{bmatrix} [-4 & -6 & 2 & 1] \\ [4 & 8 & 2 & 7] \\ [2 & 4 & -4 & 1] \\ [2 & 3 & -1 & 2]] \end{bmatrix}.$$

5. (Check) PA = LU where P is the product of the elementary permutation matrices P_{12} and P_{24}. Since P is a permutation matrix, we know that $P^T = P^{-1}$. Thus $P^T LU = P^{-1} LU = A$. Recall P to level 1 and get P^T, SWAP levels with LU and then use $\boxed{*}$ to see $P^T LU = A$.

CAS in Differential Equations and Linear Algebra

Patrick Sullivan
Valparaiso University

1 Introduction

This article will describe how a CAS is used in our sophomore course in linear algebra and differential equations at Valparaiso University. Most of the students in this course are engineering and science majors, so applications are important.

Computers were first integrated into this course in the spring of 1990 using *muMATH* and *MICRO-CALC*. In the fall of 1990 *muMATH* was replaced with *Derive*. Generally *Derive* could do everything *muMATH* could but with a vastly superior user interface. The computer is used by the instructors as a demonstration tool during class and by the students for homework assignments and take-home exams.

The main topics covered in this course are first-order differential equations, basic matrix algebra, linear differential equations, vector spaces, and linear systems of differential equations. Having the computer available has significantly altered how this class is taught.

2 First-Order Differential Equations

Probably the biggest change made in the syllabus was to substantially reduce the amount of time devoted to first-order differential equations. The only first-order equations solved in closed form are linear and separable ones. To study other first-order equations, students could use *muMATH* to solve those that had closed-form solutions (in fact *muMATH* "knew" more solution methods than the instructors) and *MICRO-CALC* to get direction fields and numerical solutions for those equations that did not have a closed-form solution. Unfortunately, when we changed to *Derive*

the one thing that was lost was the capability of finding closed-form solutions easily, but I suspect that will change. Still, I found the abandonment of most of the tedious first-order methods to be a welcome change in this course. Too often students perceive a first course in differential equations as a series of recipes. By cutting back on the different methods presented, more time can be spent studying modeling with differential equations and analyzing the solutions obtained. Another big advantage to having a CAS available is being able to do interesting problems that require nontrivial integrals without spending time in class reviewing integration by parts, trigonometric substitutions, or partial fraction decompositions.

Let me give some examples of the types of problems on first-order differential equations that were given on either take-home exams or homework assignments.

Example 1: Direction Fields.

1. Print out a direction field for the differential equation
$$\frac{dx}{dy} = \frac{9x + y}{x - 4y}.$$
   On this direction field sketch a few solutions by hand.

2. Solve the differential equation given above using *muMATH*. (Something to think about: Which solution method is more useful in understanding the differential equation?)

The above differential equation has closed-form solution (at least according to *muMATH*)

$$-6C - 6\ln x - 3\ln\left(9 + 4\frac{y^2}{x^2}\right) + \arctan\left(\frac{2y}{3x}\right) = 0.$$

The direction field shows easily that the solution curves are spirals. In the past the most we could hope for students to accomplish was to solve this differential equation in closed form by hand. Students were not likely to get the correct answer and the solution equation obtained above is not meaningful to most students. By contrast, the spiral shape of the solution curves was readily apparent to most students after finding the direction field. This problem leads to a discussion of which method is better for studying this differential equation. I try to point out the pros and cons of both the closed form and the direction field. An advantage of the closed form is that it is precise, so if numerical answers are necessary these answers will be exact. An advantage of the direction field is it shows the general behavior of the solutions. This information is hard to glean from the implicit closed-form solution even using the graphics capabilities of many programs.

Example 2: Heat Transfer. Suppose the heating system of a house could, if the house were perfectly insulated, raise the temperature $10°F$ per hour. However, the house is not perfectly insulated. It loses heat according to Newton's law of cooling.

1. If the heat is turned off when the house is at $70°F$ and the temperature outside is $30°F$, then the house temperature reaches $60°F$ after 2 hours. How long does it take, from the time the heat is turned off for the house to reach $50°F$?

2. When the temperature inside reaches $50°F$, the heat is turned back on. What will the temperature be 2 hours later?

3. When the temperature inside reaches $70°F$, the outside temperature drops steadily at $15°$ per hour. If the heating system remains on, what will the temperature of the house be when the temperature outside reaches $0°F$?

4. Over the two-hour time period from when the house is $70°$ until the outside temperature is $0°F$ what is the maximum temperature of the house to the nearest tenth of a degree?

Parts 1, 2, and 3 of this example are taken from [2, pp. 40–41 no. 17]; Part 4 was my own addition. This is a nice problem because Parts 1, 2, and 3 all require a new differential equation, but one of the parameters is the same for all three equations. Parts 1 and 2 lead to separable equations but Part 3 gives rise

to a linear non-separable equation. I would be reluctant to assign something like Part 4 without a CAS available to either get an approximation to the maximum from a graph of the function or to do the symbolic manipulations necessary to find the maximum in closed form. Moreover, it is important for students to realize the importance of analyzing their solutions to differential equations. This helps one to decide if the differential equation is a reasonable model of the situation. It also reinforces the concepts taught in previous courses. Access to a CAS puts this kind of analysis within reach of the students whereas before such tools were available, this type of analysis could be done only in the simplest examples.

Example 3: Chemical Reaction Rate. This example was a problem on the take-home portion of the first exam. The students were given a hand-out describing the differential equation necessary to model this situation and a couple of examples. They then were given the following problem.

Suppose that 2 lb of chemical A and 3 lb of chemical B react to form 4 lb of chemical C and 1 lb of a by-product, and the rate at which chemical C is produced is proportional to the amounts of A and B present.

1. Suppose that we initially have 80 lb of A, 120 lb of B and 0 lb of C and that after 1 hour 100 lb of C have been produced.

   (a) If $x(t)$ is the amount of C formed after t hours, find a formula for $x(t)$.

   (b) Show that your answer to Part (a) tends to 160 as t tends to ∞.

   (c) How long does it take to produce 150 lb of C?

Note that the answer of 160 for the limiting value of C is not surprising because of the proportions. Since 2 lb of A and 3 lb of B produce 4 lb of C multiply everything by 40! Note that if the amount of A is increased but the amount of B is not changed, the limiting value for C will still be 160 lb because that is the most that 120 lb of B can produce. However, increasing the amount of A can speed up the reaction, as we shall see.

2. Suppose we start with M lb of A ($M > 80$), 120 lb of B and 0 lb of C. Find out what M should be, to the nearest pound, so that it takes only one hour to produce 150 lb of C. Assume that the proportionality constant α is the same as in

Part 1. Suggestions: Let $x(t)$ be the amount of C at time t, set up a differential equation for $x(t)$ with M in it, separate, integrate, use the initial condition that $x(0) = 0$, use the fact that $x(1) = 150$ to get an equation for M. You will need to approximate the solutions to this equation.

Part 1 is straight-forward, but for many students using the CAS for the integration makes obtaining the correct answer more likely. However, Part 2 is more challenging. To solve the separable differential equation involved requires a partial fraction decomposition of a function with a parameter. The symbolic capabilities of a CAS are needed here because numerical solutions of either the integrals or the differential equations would be impossible because of the parameter. Then the students need to find an approximation to an equation that is impossible to solve algebraically. This requires the use of the numerical or graphical capabilities of a CAS. I think it is important that students realize the difference between an exact, symbolic computation and an approximate, numeric one. Also they need to realize that both approaches are valuable and, as this problem illustrates, both may be necessary.

3   Systems of Linear Equations

Having a CAS available is also helpful when beginning elementary matrix algebra. The students are still expected to do row reduction, matrix multiplication, determinants and inverses by hand, at least for small matrices, but they are encouraged to use the computer for larger problems.

Example 4: Temperature Gradients. In the insulated metal plate shown in Figure 1 the numbers represent temperatures at the indicated points. Assume that each interior temperature is the average of the temperatures at the four neighboring grid points.

1. We first treat the b_i's as known and the t_i's as unknown. Solve for the t_i's in terms of the b_i's.

2. Find the t_i's if $b_1 = 27$, $b_2 = 36$, $b_3 = 38$, $b_4 = 29$, $b_5 = 21$, and $b_6 = 18$.

3. We will now treat the t_i's as known and the b_i's as unknown. Rewrite the system of equations to reflect this (this results in 6 unknowns but only 4 equations).

4. Solve for the b_i's.

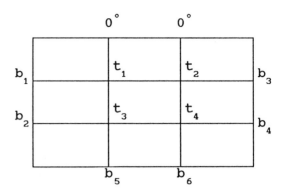

Figure 1: Temperatures on a metal plate.

5. Suppose we want $t_1 = 100$, $t_2 = 75$, $t_3 = 90$, and $t_4 = 190$. Suppose in addition we want $b_i \geq 0$ for each i and b_5 and b_6 as large as possible. Solve for the b_i's.

This is based partially on a problem from [1, p. 42 no. 10]. To solve these problems students must be able to apply row reduction to a matrix with parameters in it. If students were to attempt this by hand they would become so caught up in the row reduction that they would lose sight of their goal; by using the CAS for the computations they can focus their attention on the relationship between the two sets of variables. This problem also helps students to appreciate the distinction between unknown variables and known parameters. Also, Part 1 has a unique solution for all choices of the b_i's whereas Part 4 has infinitely many solutions for all choices of the t_i's, which helps students to distinguish between these two types of systems of equations.

Example 5: Wheatstone Bridge Circuit. This example is something I did in class. It involves the wheatstone bridge circuit (see Figure 2). The textbook [3, pp. 131–132] shows the wheatstone bridge and sets up the system of equations for the circuit. One of the exercises in the textbook [3, p. 205 no. 1] gives values for V and the R_j's and asked for the i_j's. I first decided to keep V and the R_j's as unknown parameters and used the CAS to solve for the currents. This yielded (very messy) formulas for the currents in terms of the voltage and the resistances. But I later discovered that this is not the way the circuit is actually used! It turns out that what you are given is V, R_1, R_2, and R_4 and that R_4 is chosen so that $i_3 = 0$. The problem is then to find R_5. Viewed

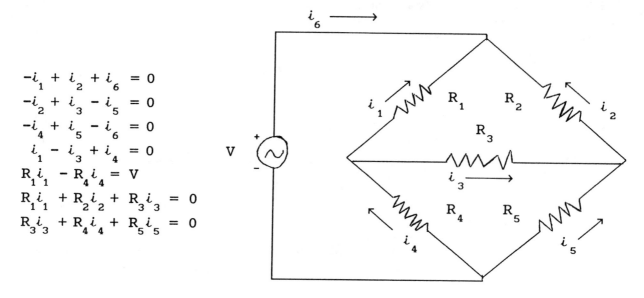

$$-i_1 + i_2 + i_6 = 0$$
$$-i_2 + i_3 - i_5 = 0$$
$$-i_4 + i_5 - i_6 = 0$$
$$i_1 - i_3 + i_4 = 0$$
$$R_1 i_1 - R_4 i_4 = V$$
$$R_1 i_1 + R_2 i_2 + R_3 i_3 = 0$$
$$R_3 i_3 + R_4 i_4 + R_5 i_5 = 0$$

Figure 2 : The wheatstone bridge circuit.

in this way the system is in fact nonlinear because of the last equation and the fact that both R_5 and i_5 are unknown. However, if you drop the last equation you have a linear system in the unknowns i_1, i_2, i_4, and i_6 (recall that $i_3 = 0$). This system of 6 equations and 5 unknowns has a unique solution for all choices of V, R_1, R_2, and R_4. You can then use the solution for i_4 and i_5 in the last equation to solve for R_5. This example points out the necessity of using matrix techniques to solve linear equations. It also shows that sometimes adjustments need to be made in problems to apply our mathematical knowledge. Again, the linear systems here are very difficult to do by hand even if the values of all the parameters are specified. But a CAS makes finding the solutions even with the parameters unspecified a relatively easy task. The hard part is knowing which equations to solve. By relegating most of the tedious row-reduction calculations to the CAS, students can look at more examples of systems of equations and begin to acquire the experience necessary to know which equations to solve.

4   Higher Order Linear Differential Equations

The next topic covered is linear differential equations of second and higher orders. A little of the the-

ory is discussed (superposition principle, the Wronskian, etc.) then constant coefficient problems and the method of undetermined coefficients for nonhomogeneous problems are covered. A CAS is helpful for computing Wronskians, finding roots of characteristic polynomials and finding the necessary derivatives and solving the systems of equations arising in undetermined coefficient problems.

One of the applications of this subject is damped oscillators. Let me give some examples of the types of problems given to the students.

Example 6: Motion of a Spring.

1. Find the general solution of the o.d.e. modeling a damped spring with mass 5 g, spring constant 60 dynes/cm and damping constant 40 dynes/(cm/sec).

2. In Part 1 what is the limit as $t \to \infty$?

3. Suppose we start with the spring unstretched and an initial velocity of 3 cm/sec. What is the maximum the spring is stretched?

4. Suppose the spring can only be stretched 12 cm before it breaks. If the spring begins unstretched what is the largest value of the initial velocity that will not break the spring?

This example requires the student to analyze the solution obtained from the differential equation. In Part 3 this analysis is easily accomplished either graphically or by using the first derivative to locate the maximum. It is necessary in Part 4, however, to find the maximum of a function with an unknown parameter (the initial velocity). The students are then forced to use symbolic computations to solve the problem. Again, having a CAS do these computations helps students to understand what the solutions of the differential equation tell them about the properties of this system. One interesting feature of this problem is that the time necessary for the spring to reach its maximum is independent of the initial velocity.

Example 7: A Floating Box. This example begins with a problem from [2 pp. 46–47 no. 1].

1. A cube of wood 1 foot on each side, weighing 16 lb (so that its mass is 16/32 slugs), floats in a river with the bottom of the box x feet $(x < 1)$ below the surface of the water. According to the Principle of Archimedes the water buoys up the box by a force equal to the weight of the water it displaces. Given that the density of the water is 62.5 lb/ft^3, write an o.d.e. for x assuming resistance with damping constant b lb/(ft/sec).

2. Solve the o.d.e. from Part 1 when $b = 10$.

3. I want to look at the answer to Part 2,

$$x = c_1 e^{-10t} \cos(5t) + c_2 e^{-10t} \sin(5t) + \frac{32}{125},$$

in more detail. Note that if $x > 1$ then the box is completely submerged and the solution to the differential equation no longer gives the motion of the box.

   (a) Suppose we start with the box on the surface of the water $(x = 0)$ and initial velocity, v, downward (which is the positive direction). Solve for c_1 and c_2 in terms of v.

   (b) We wish to find the largest velocity which will not swamp the box. Solving for v exactly is a difficult problem. Instead, plug in a few values for v and graph the resulting functions you get for x. Use this to approximate within 1 ft/sec the largest velocity which does not swamp the box.

4. Take the solution to Part 2 but with viscous damping with $b = 1$.

   (a) Show that the general solution is given by

   $$x = c_1 e^{-t} \cos(2\sqrt{31}t) + c_2 e^{-t} \sin(2\sqrt{31}t) + \frac{32}{125}.$$

   (b) Suppose we start with the box on the surface of the water $(x = 0)$ and initial velocity, v, downward (which is the positive direction). Solve for c_1 and c_2 in terms of v.

   (c) Our solution is only valid if $0 < x < 1$. Again, pick a few values of v and graph the resulting functions you get for x. Use these graphs to approximate within 0.1 ft/sec the largest velocity for which $0 < x < 1$ for all time.

These problems deal mainly with an analysis of the solutions of some differential equations. These problems demonstrate some of the limitations that differential equations models may have, namely that our mathematical solutions are not always physically meaningful. In Part 3, if the initial velocity is too large the box is swamped and the differential equation is no longer valid. In Part 4, because of the lowered damping constant, if the initial velocity is too large, the solution of the differential equation predicts that the box is flung out of the water at some time and again the differential equation (and our solution) no longer apply. These examples help students to understand the effects of damping on this system. The damping term, instead of just being a mysterious $b\frac{dx}{dt}$ term in the differential equation, becomes physically meaningful. Also these problems allow a discussion of how reasonable are the mathematical solutions to this problem and how well does this o.d.e. model a floating box. To actually find the limiting velocity in each case, experimenting with graphs and different values of v is about the only practical method. To do this successfully without a CAS would be impossible for even the most gifted student.

Example 8: Spring with Damping. This example is from a take-home exam.

We have a damped mass-spring system, with mass 1 g, spring constant 2 dynes/cm and an unknown damping constant. We stretch the spring 2 cm and release (so the initial velocity is 0). Two seconds later the spring crosses the equilibrium position (i.e., $x = 0$). Find an approximation of the damping constant good to two decimal places. Give the proper units.

In this question the students must analyze the solution to the damped spring given in the text. It turns out that the equation they must solve to get the damping constant is impossible to solve algebraically so some sort of approximation technique is necessary. I like this problem because a damping constant is usually determined empirically instead of theoretically and here is a way to use physical observations to determine the damping constant. Like most mathematicians I have never actually tried to do this with a physical system.

5 Vectors in $\Re^n$

The next topic studied is vector spaces. The main emphasis is to get the students to understand the concepts of subspace, spanning, linear independence, and bases at least in $\Re^n$. Proofs are not emphasized; the students are expected to be able to give examples of these concepts and to apply these concepts to concrete examples, usually in $\Re^n$. The main difficulty with working with specific examples is that the computations are time-consuming and difficult to do accurately by hand. The computer helps to ease some of these problems.

Probably one of the more difficult ideas to get across is spanning. For example, if one wishes to show that the vectors $[1\ 0\ 1]^T$, $[1\ 1\ -1]^T$, $[2\ 1\ 0]^T$, and $[1\ 0\ 0]^T$ span $\Re^3$ then one must row-reduce the matrix

$$\begin{bmatrix} 1 & 1 & 2 & 1 & x \\ 0 & 1 & 1 & 0 & y \\ 1 & -1 & 0 & 0 & z \end{bmatrix}.$$

Having a CAS available greatly increases the chance that this row-reduction will be done successfully, and more effort can be put into understanding the results of the row-reduction. Similarly, testing for linear independence frequently requires row-reduction of a matrix. Having the CAS available enables one to look at more examples to reinforce these concepts.

6 Linear Systems of Differential Equations

When studying systems of differential equations we write them in matrix form and find solutions using eigenvalues, eigenvectors, and generalized eigenvectors. Students are still expected to be able to find eigenvalues and eigenvectors by hand, at least for

small matrices. But having the CAS available for these sorts of computations is invaluable. One can actually solve 3×3 or higher systems without using up a whole class period on one problem.

Example 9: The Effect of the Multiplicity of Eigenvalues. This problem was assigned to give students some feel for what the multiplicity of an eigenvalue in the characteristic equation implies about eigenspaces and generalized eigenspaces.

The characteristic equation of the matrices, A, that follow is $(\lambda-2)^3(\lambda-1)$. (You may want to check this.) For each matrix find a basis for the

1. Eigenspace of A corresponding to the eigenvalue 2;

2. Solution space of $(A-2I)^2\vec{v} = \vec{0}$;

3. Solution space of $(A-2I)^2\vec{v} = \vec{0}$ that includes the vectors from Part 1 (this may be the same as Part 2);

4. Solution space of $(A-2I)^3\vec{v} = \vec{0}$;

5. Solution space of $(A-2I)^3\vec{v} = \vec{0}$ that includes the vectors from Part 3 (this may be the same as the space in Part 4); (the solution space of $(A-2I)^n\vec{v} = \vec{0}$ will be the same as the solution space of $(A-2I)^3\vec{v} = \vec{0}$ for all $n > 3$);

6. Eigenspace of A corresponding to the eigenvalue 1;

7. Solution space of $(A-I)^2\vec{v} = \vec{0}$ (should be the same as the space in Part 6).

$$A = \begin{bmatrix} 1 & -1 & 0 & 1 \\ -2 & 0 & 0 & 2 \\ -1 & -1 & 2 & 1 \\ -2 & -2 & 0 & 4 \end{bmatrix},$$

$$A = \begin{bmatrix} 4 & 4 & 4 & 4 \\ -1 & 0 & -2 & -2 \\ -3 & -5 & -3 & -4 \\ 3 & 5 & 5 & 6 \end{bmatrix},$$

$$A = \begin{bmatrix} 1 & -1 & 0 & 2 \\ -1 & 0 & -1 & 4 \\ -1 & -1 & 1 & 3 \\ -1 & -1 & -1 & 5 \end{bmatrix}.$$

The first matrix in this example has a three-dimensional eigenspace corresponding to the eigenvalue 2, the second matrix has a two-dimensional eigenspace corresponding to 2, and the third has a one-dimensional eigenspace corresponding to 2. I would never have given an assignment like this without a CAS available to do these calculations. The main point of these exercises is to show how the dimension of the eigenspace is less than or equal to the multiplicity of the corresponding eigenvalue, and that there is a generalized eigenspace of the same dimension as the multiplicity of the eigenvalue. This will be useful when we are looking for complete sets of solutions to systems of differential equations.

Example 10: Radioactive Decay. This and the next two examples are applications of systems of differential equations. This example is based on some problems from [2, p. 145 no. 2 and pp. 260–261 no. 19].

A radioactive substance decays at the rate of 50% annually into substance B. Substance B decays at the rate of 10% annually into substance C. Suppose that α grams per year of substance A are steadily added to the mixture while substance C is extracted steadily at the rate of 100 grams per year, and initially there are 500 grams of each substance. Assume that each gram of A decays into one gram of B and each gram of B decays into one gram of C.

1. Find a system of differential equations to model this situation.

2. Find formulas for the amount of substances A, B, and C at time t.

3. Find the smallest value of α so that the amounts of substances A, B, and C never reach 0.

In the original text-book version of this problem, α is given the value of 250 and the solver is asked whether this is large enough to prevent the amounts of substance A, B, and C from ever reaching 0. To solve this non-homogeneous system of differential equations I used variation of parameters in the following form. To solve a system of the form

$$D\vec{x} = A\vec{x} + \vec{E}(t),$$

where A is a known $n \times n$ constant matrix, $\vec{E}(t)$ is a known $n \times 1$ matrix of functions and $\vec{x}$ is an unknown $n \times 1$ matrix of functions, one first solves the related homogeneous system $D\vec{x} = A\vec{x}$. Using the eigenvalues and eigenvectors of A, one constructs $\Phi(t)$, a fundamental matrix of the homogeneous system. Then a particular solution for the original non-homogeneous

problem is

$$\Phi(t) \int \Phi^{-1}(t)\vec{E}(t)\,dt.$$

These calculations are done very easily on a CAS, even with α as an unknown parameter. After we have found formulas for the amounts of substances A, B, and C as functions of time, the analysis of these functions to determine the appropriate value of α is greatly aided by the CAS. The formulas are:

$$2\alpha - 2(\alpha - 250)e^{-0.5t},$$

$$2.5(\alpha - 250)e^{-0.5t} - 12.5(\alpha - 90)e^{-0.1t} + 10\alpha,$$

and

$$12.5(\alpha - 90)e^{-0.1t} - 0.5(\alpha - 250)e^{-0.5t}$$

$$+ (\alpha - 100)t - 12\alpha + 1500$$

for substances A, B, and C, respectively. It is easily seen that the amount of C will be negative for large t if α is less than 100. This leads one to conjecture that $\alpha = 100$ is the smallest amount that will work. Graphing all three functions for this value of α supports this conclusion. Finally, one can verify that this is correct by looking at the minimum value of each of the above formulas and seeing how it depends on α. This example requires both the use of a parameter as well as an analysis of the solution of the system.

Example 11: A Mixing Problem. This example is from a take-home exam and is based on an example from [4, p. 252 example 1].

1. Tank X contains 50 gallons of water in which 12 lb of salt are dissolved and Tank Y contains 50 gallons of water with no salt. Water containing no salt flows into X at 8 gal/min; the mixture (well-stirred) in X flows from X to Y at 9 gal/min; the mixture in Y (also well-stirred) flows back into X at 1 gal/min and out of the system at 8 gal/min. Let $x(t)$ and $y(t)$ denote the amounts of salt in pounds in tanks X and Y at time t.

   (a) Find $x(t)$ and $y(t)$.

   (b) Find the limit as $t \to \infty$ of $x(t)$ and $y(t)$.

   (c) For what values of t is $x(t) > y(t)$?

   (d) For what values of t is $x(t)$ increasing?

   (e) For what values of t is $y(t)$ increasing?

2. We have the same set-up as in Part 1, except that the water flowing into tank X at 8 gal/min has 1 lb/gal of salt in it. Repeat Parts 1(a) through 1(e).

The main computational difficulties in this problem are finding the necessary eigenvalues, performing variation of parameters to solve the non-homogeneous system which arises in Part 2, taking the necessary derivatives for solving Parts 1(d) and 1(e) of both problems and solving the necessary equations for Parts 1(c), 1(d), and 1(e) for both problems. Having the CAS available is helpful for all these steps.

When the students are given the solutions to these problems, I like to talk about how some of the answers can be anticipated. For example, a little thought before solving the system of differential equations tells you what the limiting values of x and y must be in Parts 1 and 2. In both cases the limiting concentration of salt in each tank must be the same as the concentration of salt in the water flowing into tank X from outside at 8 gal/min. So in Part 1, the limit of x and y must be 0 and in Part 2, the limit of x and y must be 50. The solutions of the systems of differential equations confirm this. The other fact that can be anticipated is that for both problems the answer to Part 1(c) must equal the answer to Part 1(e). To see this, look at the differential equation for tank Y (which is the same for Parts 1 and 2),

$$\frac{dy}{dt} = \frac{9}{50}(x - y).$$

It is now obvious that $x(t) > y(t)$ if and only if $y(t)$ is an increasing function.

7   Conclusions

In the examples presented in this paper there are some common themes. Many of these problems are not that different from the ones appearing in standard textbooks. However, they have been pushed beyond what we traditionally did by hand. Ideally, the problems become more realistic to the engineering and science majors who take this course. But, they also learn some new mathematics as well as become more proficient at previously learned mathematics.

Probably the most common alterations that have been made in these problems are the insertions of unknown parameters for known fixed constants and the analysis of the solutions of the differential equations.

I have shown several different problems in differential equations and linear algebra where using parameters is more logical than using fixed constants. There are two main approaches one can take in problems with parameters: find a closed form solution involving the parameters or experiment with several different fixed values of the parameters. A CAS is useful for both approaches.

Finding the solution to a differential equation is a lot of work. In the past, if you were then to ask the students to analyze their solution, perhaps using techniques from previous semesters of calculus, you would probably have a revolt on your hands. However, now it is quite easy to use the computer to analyze these solutions.

The students can graph their solutions, find extreme points, or discover what happens to their solutions in the long term. After solving a system of differential equations they can compare the solutions of the dependent variables. This analysis helps begin a discussion of the validity of the differential equation models that are being used.

This course has always had too much material in it. The engineering and science majors in the course always wanted more applications and less theory. The CAS helped me to trim some of the material and also give the students more problems involving applications.

References

1. Cullen, C. G. (1988). *Linear Algebra with Applications*, Scott, Foresman and Co., Glenview, IL.

2. Nitecki, Z. H. and Guterman, M. M. (1986). *Differential Equations with Linear Algebra*, Saunders College Publishing, Philadelphia, PA.

3. Powers, D. L. (1986). *Elementary Differential Equations with Linear Algebra*, Prindle, Weber and Schmidt, Boston, MA.

4. Tierney, J. A. (1985). *Differential Equations*, 2nd Ed., Allyn and Bacon, Inc., Boston, MA.

Linear Systems of Differential Equations
Via *Maple*

Robert J. Lopez
Rose-Hulman Institute of Technology

1 Introduction

An early teaching assignment, after coming to Rose-Hulman Institute of Technology in 1985, was to teach the second quarter of ordinary differential equations to two sections of engineering and science students. This course typically began with the study of linear systems of ordinary differential equations. Computations usually were done in the case $n = 2$, and, sometimes, in the case $n = 3$.

The recommended syllabus had the students first solve such systems by differentiation and elimination, then by Laplace transform (which they had learned the quarter before), and finally by eigenvector methods. After two tries at this syllabus, first in 1985, then in 1987, I decided that without appropriate computational tools, the syllabus could not be mastered by any but the most algebraically adept students.

In June 1988 a $100,000 Instrumentation and Laboratory Improvement (ILI) proposal to establish a computerized classroom for making computer algebra the tool of first recourse in teaching, learning, and doing engineering analysis at Rose-Hulman Institute of Technology was funded by NSF. This classroom, called Symlab, was opened in January 1989. It was used for the first time in teaching linear systems of differential equations in the winter quarter of 1989.

The results of these initiatives with *Maple* have been encouraging. In comparison to traditionally taught students from previous terms, a greater percentage of the *Maple* students were found to learn the central ideas and to master the calculations. More time was spent formulating problems and interpreting solutions; the unit ceased being a month of drill in algorithms and became a true interactive learning experience.

We present, then, a sketch of how *Maple* has transformed an otherwise formidable topic into some living mathematics.

2 Motivation

Because the unit on linear systems occurs in our second quarter of differential equations, the students have already seen Laplace transforms and simple mixing problems of the form

> A 300 gallon tank contains 100 gallons of brine at a concentration of 30%. A 50% brine solution enters the tank (where it mixes instantly with the existing contents) at a rate of 25 gallons per minute. What is the concentration of salt in the tank when the tank overflows?

Hence, it is possible to use multi-tank mixing problems to generate linear systems of ordinary differential equations. For example, we pose the following problem.

> Tank A initially holds 50 gallons of pure water, tank B, 100 gallons of brine at a concentration of 3/4 pound per gallon. Instantaneous mixing occurs when liquid is pumped back and forth between the tanks at a rate of 5 gallons per minute. Find the amounts of salt in each tank as a function of time.

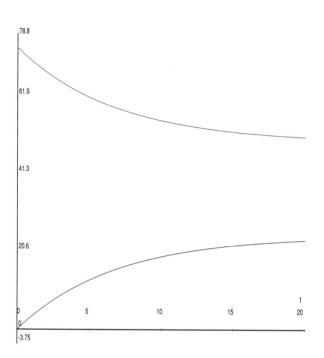

Figure 1: Graphs of $x(t)$ and $y(t)$.

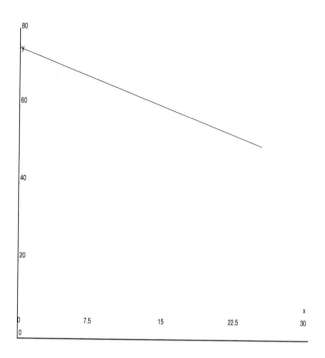

Figure 2: System trajectory given by $y = y(x)$.

If $x(t)$ and $y(t)$ represent the amounts of salt in tanks A and B respectively, then mass balance leads to the following linear system

$$\frac{dx}{dt} = -\frac{5}{50}x + \frac{5}{100}y$$
$$\frac{dy}{dt} = \frac{5}{50}x - \frac{5}{100}y \qquad (1)$$

with $x(0) = 0$ and $y(0) = 75$ as initial conditions.

There are at least three approaches that can be taken to solve this system. What is important, however, is that the system be solved promptly and the solution studied, in order to make clear the nature of the problem and its solution. Thus, system (1) can be solved directly by the *Maple* `dsolve` command, by taking Laplace transforms and implementing the appropriate algebra in *Maple*, or by directly entering the transformed version of (1) into *Maple* and using *Maple* to finish extracting the solution. We illustrate the third alternative, using $X(s)$ and $Y(s)$ to denote the transforms of $x(t)$ and $y(t)$ respectively. Thus,

$$sX = -X/10 + Y/20$$
$$sY - 75 = X/10 - Y/20 \qquad (2)$$

A call to the *Maple* `solve` command returns

$$X(s) = \frac{75}{s(20s + 3)}$$
$$Y(s) = \frac{75(20s + 2)}{s(20s + 3)} \qquad (3)$$

from which the *Maple* `invlaplace` command inverts each transform to

$$x(t) = 25(1 - e^{-3t/20})$$
$$y(t) = 25(2 + e^{-3t/20}) \qquad (4)$$

In the traditional approach to obtaining this solution, the manipulations themselves become the message. With *Maple*, this is not the case. In addition, *Maple* allows us to examine the solution for its relation to the problem that generated it. Thus, in Figure 1 we see evidence that both $x(t)$ and $y(t)$ might be tending toward limits. This is confirmed analytically by invoking the *Maple* `limit` command to find that $x(t) \rightarrow 25$ and $y(t) \rightarrow 50$ as t gets arbitrarily large.

Moreover, students familiar with the notion of the parametric form of a curve can obtain the graph of $y(x)$ via *Maple*'s parametric plot facility. In Figure 2 we see the result of the *Maple* command `plot([x(t),y(t),0..50])`. The trajectory $y(x)$ is one path in the phase plane. It is now natural to discuss and explore the notion of the phase plane.

In this example the graph of the system trajectory appears to be a straight line. There are two ways that this can be verified. First, the parameter t can be eliminated from (4) by adding $x(t)$ and $y(t)$. Inspection shows that $x + y = 75$, the equation for a straight line.

Alternatively, the differential equation

$$\frac{dy}{dx} = \frac{dy/dt}{dx/dt} = \frac{x/10 - y/20}{-x/10 + y/20} = -1 \qquad (5)$$

along with the initial condition $y(0) = 75$, is readily solved for the same linear solution $x + y = 75$. Both approaches demonstrate fundamental principles, the first being that t is a parameter on individual paths, and the second, that the paths are solutions of a single ordinary differential equation.

3 Further Example

It is interesting to formulate a slightly more complex "mixing-tank" problem for which the solution will present greater computational complexity. Consider, for example, the following problem.

Tank A contains 100 gallons of brine at a concentration of 1 pound of salt per gallon. Tank B contains 100 gallons of pure water. Water flows into tank A at a rate of 2 gallons per minute, liquid flows from tank A to tank B at a rate of 3 gallons per minute, liquid flows from tank B to tank A at the rate of 1 gallon per minute, and liquid overflows from tank B at a rate of 2 gallons per minute. Assuming instantaneous mixing in both tanks, find the amounts of salt in tanks A and B as a function of time t.

The linear differential equations defining this system are

$$\begin{aligned} \frac{dx}{dt} &= -\frac{3}{100}x + \frac{1}{100}y \\ \frac{dy}{dt} &= \frac{3}{100}x - \frac{3}{100}y \end{aligned} \qquad (6)$$

along with initial conditions $x(0) = 100$ and $y(0) = 0$, provided $x(t)$ and $y(t)$ are defined as the amounts of salt in tanks A and B respectively.

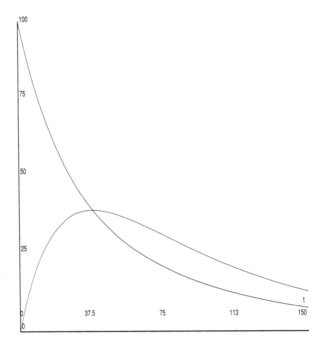

Figure 3: Graphs of $x(t)$ and $y(t)$.

Using *Maple* to generate the solution, whether by Laplace transforms or by the *Maple* `dsolve` command, leads to

$$\begin{aligned} x(t) &= 50e^{(\sqrt{3}-3)t/100} + 50e^{-(\sqrt{3}+3)t/100} \\ y(t) &= \frac{150}{\sqrt{3}}e^{(\sqrt{3}-3)t/100} - \frac{150}{\sqrt{3}}e^{-(\sqrt{3}-3)t/100} \end{aligned} \qquad (7)$$

whose graphs are shown in Figure 3. The graphs of $x(t)$ and $y(t)$ suggest that the amount of salt in each tank is approaching zero as time increases without bound. An appeal to the *Maple* `limit` command again confirms this behavior. In fact, the student is led to notice that the coefficient of t in each exponential appearing in (7) is negative. Hence, the limiting behavior is immediate.

However, when the graph of $y(x)$ is obtained from *Maple*'s parametric plot facility, as in Figure 4, the more complex behavior of the system trajectory is observed. The rise and fall of the amount of salt in tank B is not immediately suspected and would not be apparent without this plot!

Since the system trajectory in the phase plane is not the straight line of the first example, we can expect greater complexity when trying to obtain an explicit expression for the path $y(x)$. If we write (7) in the

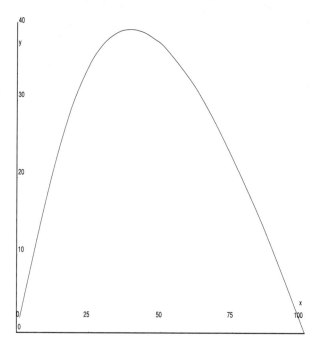

Figure 4: System trajectory of $y = y(x)$.

form

$$x(t) = 100 e^{-3t/100} \cosh \frac{\sqrt{3}}{100} t$$

$$y(t) = 100\sqrt{3} e^{-3t/100} \sinh \frac{\sqrt{3}}{100} t$$

then

$$\frac{y}{x} = \sqrt{3} \tanh \frac{\sqrt{3}}{100} t$$

so that

$$t = \frac{100}{\sqrt{3}} \operatorname{arctanh} \frac{y}{\sqrt{3}x}. \tag{8}$$

Thus,

$$3x^2 - y^2 = 30000 \exp\left(-2\sqrt{3} \operatorname{arctanh} \frac{y}{\sqrt{3}y}\right). \tag{9}$$

Finally, it is instructive to verify that the function $y(x)$ defined implicitly by (9) is, in fact, the same function defined by (7). Clearly, this has to be done numerically, and we now illustrate how *Maple* allows us to shift focus from analytic results to numeric results.

To this end we will assign x a numeric value in (9) and numerically compute the corresponding $y(x)$. Figure 5 shows a table of values so computed, while

k	x_k	$y_k = y(x_k)$
1	10	16.9124129477
2	20	30.2116192162
3	30	37.3431259072
4	40	39.1356339842
5	50	37.1742993417
6	60	32.6219404382
7	70	26.2232869304
8	80	18.4572441635
9	90	9.6426356121

Figure 5: Selected points (x, y) computed from (9).

Figure 6 shows the smooth curve *Maple* has put through the data points of Figure 5. Comparing Figures 5 and 6 suggests that $y(x)$, as defined by either (7) or (9), is the same function. This can be further confirmed by solving $x(t) = 10k$, $k = 1, ..., 9$ in (7) for t, and substituting each computed value of t into $y = y(t)$ in (7). That the results of this computation agree with the values in Figure 5 is not surprising.

The reader should not conclude that any particular calculation described above is of such unique importance that the calculations themselves are the content of this discussion. The shrewd reader will understand that the pedagogic activities just described make the abstractions of concepts like the phase plane and system trajectories "real" to students. There may well be many other learning activities even better suited to these ends than any presented here. The author is a willing learner!

4 Vectors

Finally, let's look at how having the solutions in hand motivates the search for vectorial solution techniques. Begin by inspecting the solutions (4) and (7). Especially from (7) we can see that the solution naturally decomposes into the vector format

$$\begin{bmatrix} x \\ y \end{bmatrix} = 50 \begin{bmatrix} 1 \\ 3 \end{bmatrix} e^{(\sqrt{3}-3)t/100}$$

$$+ 50 \begin{bmatrix} 1 \\ -3 \end{bmatrix} e^{-(\sqrt{3}+3)t/100}.$$

The next step is to write the system (6) as

$$\begin{bmatrix} x \\ y \end{bmatrix}' = \begin{bmatrix} -3/100 & 1/100 \\ 3/100 & -3/100 \end{bmatrix} \begin{bmatrix} x \\ y \end{bmatrix} \tag{10}$$

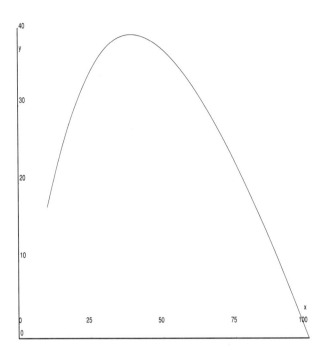

Figure 6: *Maple* spline fit to computed (x, y) pairs.

and to recognize the structure $\mathbf{X}' = \mathrm{A}\,\mathbf{X}$ therein contained. If the solution to this system of differential equations is going to have a vector nature, then perhaps we should expect the form to be $\mathbf{B}\,e^{kt}$ with $\mathbf{B}$ a constant column vector. That the solution should be an exponential follows from an appeal to solutions such as (4) or (7) and from an appeal to the first order equation $y' = ay$, whose solution is $y = ce^{at}$.

It should now be obvious how to proceed. Substitution of the assumed solution into (10) leads to the eigenvalue problem

$$\mathrm{A}\,\mathbf{B} = \lambda\,\mathbf{B}.$$

However, we leave it to the reader to explore the use of a computer algebra system, such as *Maple*, in solving the eigenvalue problem. For the case $n = 2$, both the eigenvalues and the eigenvectors can be obtained symbolically. However, for large systems ($n > 4$) there are no algorithms for obtaining the eigenvalues in closed form. Consequently, numerical techniques must also be included for a complete solution to the eigenvalue problem.

5   Conclusion

An often heard remark in faculty and department discussions about pedagogy at Rose-Hulman Institute of Technology (and, I would venture, at many other schools nationwide) is that the individual topics that we teach in our mathematics courses are not as important as the development of a mathematical way of thinking. My own mathematical way of thinking centers on a tinkerer's approach to a new situation. Pick up any new objects, examine them to become familiar with their properties, touch them, play with them, until the objects become familiar. Then, look for relationships of interest and explore ways of verifying those relationships.

The use of *Maple* in teaching the introduction to linear systems of differential equations, as presented above, is an outgrowth of this practical approach to grappling with new mathematics. And every time I explain why I am led to this perspective, I find it useful to relate the following.

I had gone to college and majored in mathematics, with the view of becoming a high school math teacher. My first posting had me teaching three classes of grade ten geometry. After about a month of teaching about implications, about if-then statements, and about the use of Venn diagrams to capture the content of conditionals, I found my high school students bored, lost, and dispirited. I thereupon had occasion to visit with the president of the college from which I had graduated. (He was a Ph.D. in analysis from New York University's Courant Institute.) After describing to him the situation in which I found myself I received the shocking therapy of being sharply reprimanded: "Didn't you learn anything while you were here? People learn from the particular to the general, not the other way around!"

The rest of my teaching career has been defined by my struggle to avoid making that same error again. I hope the above approach to teaching systems of linear equations shows that I have, at last, exonerated myself!

The Use of Programmable Graphics Calculators in a Differential Equations Course

T. G. Proctor
Clemson University

1 Introduction

Many universities require their engineering and science students to take three semesters of calculus followed by a course in ordinary differential equations. At Clemson, the four-credit differential equations course is usually taken at the end of the second year and requires students to construct and solve elementary mathematical models from the physical and biological sciences. The course serves as a capstone for the calculus sequence, and an attempt is made to challenge students to extrapolate their knowledge of the calculus to address a number of scientific problems new to them.

My use of technological tools in my differential equations course has changed the course significantly. What are the positive aspects to be gained from this enhancement? What are the pitfalls? What technological tools are most appropriate? I will try to answer these questions in this article: some by simple statement, some by example. The overall change in the course has been a shift in emphasis away from manual algebraic and calculus manipulations toward the use of interactive programs and, most importantly, toward the subsequent analysis of the results. A related report is given in [1].

2 Recurring Themes

The following themes occur throughout the presentation of the course:

1. The use of graphs to study families of solutions of an initial value problem resulting from different initial conditions;

2. The use of graphs to compare solutions of initial value problems resulting from two or more different mathematical models of a physical problem;

3. The determination of distinguished solutions to initial value problems such as equilibrium and periodic solutions;

4. The determination of the asymptotic behavior of solutions, especially those near a distinguished solution;

5. The use of observations of the solution to determine problem parameters;

6. The determination of the output solution function resulting from the presence of an input forcing function.

Elementary functions (exponential, polynomial, and trigonometric functions) are featured in the study of elementary differential equations. Here, students learn to recognize graphs of simple combinations of these functions. The most common combinations are formed by addition or multiplication. However, another powerful combination is the splicing together of two or more exponential, polynomial, or trigonometric functions to create a new smooth function with desired graphical properties. A tool that students can use quickly and easily to produce such graphs greatly enhances their learning.

Careless mistakes in calculation can transform the solution of a problem into garbage. A tool that helps a student avoid some of these errors and thus advance into further aspects of the problem under consideration allows the instructor to place greater emphasis on concepts and problem algorithms. Student knowledge and skills which are hidden by the use of such a tool can be easily checked by further questions. In fact, we have the opportunity to require students to go through a check drill to verify that results produced by using technology are meaningful.

Mathematical models for a selected set of physical problems are part of the traditional course in differential equations. Technology provides a tool to investigate the results of some modelling assumptions quickly. For example, frequently studied models for population growth are the logistic model,

$$\frac{dp}{dt} = -rp(1 - p/K)$$

and the Gompertz model,

$$\frac{dp}{dt} = rp\ln(K/p).$$

After analytical solutions for an initial value problem are found, a study of the results shows different asymptotic properties of the solutions and also that for sufficiently small $p(0)$, we have inflection points at $p = K/2$ and $K/2.72$, respectively. Such features may be detected first by a graph of the solutions and then verified analytically.

Next the class considers what modifications may be made to the basic models to produce different properties. Suppose that data suggests that inflection points should appear at 75% of the limiting value of the population. By creating a function F which satisfies $F(0) = F(K) = 0$ with $F(p) > 0$ for $0 < p < K$ and with $F'(.75K) = 0$, we have a model with the desired property. This can be accomplished by smoothly splicing two functions, each quadratic in p, at $p = .75K$. Even though the resulting model is analytically tractable, a graphical simulation of the solution visually makes the point and allows the class to go on with other material. In my classes such examples stimulate the class to ask many basic questions about mathematical models and to further develop their skills in studying properties of functions.

Nonlinear problems are considered as well as linear problems, especially for first- and second-order differential equations. Traditionally a great deal of attention is given to the analytical structure of the solutions of linear problems with somewhat less attention given to the resulting graphical properties. For example, transient and steady states are present in the solutions of many problems studied in the course. These properties become evident when families of solutions

arising from nearby initial conditions are graphed. Although the solution structure of nonlinear problems may be difficult to understand, it is sometimes possible to calculate special solutions (such as equilibrium or periodic solutions) and then relate other solutions to these "guideposts." The linearized version of the system near an equilibrium solution may also give important information: for example, the asymptotic approach toward limiting behavior. My students learn to look for special solutions, asymptotic properties, etc. by studying the simulations given by their calculator programs.

The need to solve several linear or nonlinear equations in as many unknowns arises naturally in this course. An example is the choice of the values of parameters in some of the mathematical models so that the solution of a differential equation appropriately fits observed data. Because such problems are computationally intensive, technology is needed to produce answers to non-contrived problems. Symbol manipulating capability is particularly useful in the nonlinear case (see below). Linear systems of differential equations are treated using traditional methods supplemented with technology for producing matrix eigenvalues and their associated eigenvectors.

3   Choice of Appropriate Technology

At our university the number of students taking all sections of this differential equations course is large. Consequently, it is impractical to purchase and maintain microcomputers and robust mathematical software to accommodate them on a regular basis. Furthermore students have little time for more laboratory work. The primary technology used in my class is a Hewlett-Packard graphic programmable calculator. Its portability allows the students to interact with the material both in class and in their study areas (library, dormitory, etc.). Extensive programming skills are not required. Most of the programming involves the development of macro programs and our workbook [2] gives examples for these macros. Only a small fraction of the students have difficulty with the programming that is required for these calculators, and it seems that delays in addressing the homework problems are the key reason for this type of difficulty.

The HP-48SX, HP-48S and HP-28S calculators have 32 kilobytes of memory (expandable on the 48SX). The programs for the course require about 10 kilobytes and, of course, the calculators are used in the students' other science and engineering courses. The calculators are durable and will usually survive a drop from desktop to floor. When the cost is amortized over the first two years in science and engineering courses, it compares favorably with software required to perform the same tasks.

The presence of calculators in students' hands in class has generated a change in my classroom presentations. I sometimes use an overhead projector to project a large image of the calculator screen: many of my students choose to follow or anticipate my steps on their own calculators. Indeed, I sometimes use the calculator as a substitute for student blackboard work. There is more than the usual amount of class discussion because the in-class problems involve more modelling and analysis.

I put questions on each of the in-class tests and on the final exam that require straightforward use of the calculator programs or the graphing capability of the calculator. Two student projects are also assigned each term which require a somewhat higher level of calculator use, and students may work on these projects in teams. Regular homework (one graded assignment on non-test weeks) may or may not include calculator problems.

Since science and engineering students will use scientific software in various settings, I discuss both the intelligent and non-intelligent use of such tools. (An example is the solution of a linear system $Ax = b$ with A "nearly" singular.) However, as this excursion from differential equations takes class time, I try to hold such discussions to five-minute explanation bytes. We also encourage students to take additional courses in scientific computation.

In summary, the calculators enable my students to spend more time in problem analysis, to avoid tedious calculations (for example, those associated with matrix manipulations), and to produce quick and, hopefully, correct graphs of functions expressible in terms of sums, products, and compositions of elementary functions.

4 First- and Second-Order Initial Value Problems

Elementary numerical techniques for the first-order initial value problem are derived and associated programs created and saved in the calculator memory early in the course. In class we restrict ourselves to the Euler and improved Euler methods rather than drift into the time sink of exploring higher order methods (which are treated in a later course). A program to produce a graph of the resulting solution or family of solutions is also created and saved so that we may study the sensitivity of the solution to initial conditions or to the parameters in the problem. For a particular problem, the students go through a checklist of requirements: a prescription of step size, number of steps, desired horizontal and vertical plot ranges, and the construction of a macro program to produce from inputs t and x, the values of the function $f(t, x)$ for the differential equation $dx/dt = f(t, x)$. Finally, they supply initial conditions t_0, and x_0 as inputs to the program which creates a graph of the approximate solution. This last step can be repeated for other initial conditions, and graphs of the solutions can be accumulated in one graph or isolated accordingly. A change in the $f(t, x)$ function can be used to compare the solutions from two mathematical models for a problem in one graph.

We previously mentioned the emphasis given to the construction of mathematical models in the course. Part of this emphasis is for the engineering students in the course, but part is to study the properties of solutions to one or more differential equations and to begin the abstraction process for general classes of problems. As an example, consider the position of a falling body subject to a resistance force opposing the fall. Some of the models, say the linear model

$$\frac{dv}{dt} = \frac{g}{m} - rv,$$

or the quadratic model

$$\frac{dv}{dt} = \frac{g}{m} - rv^2,$$

are analytically tractable. Others, such as

$$\frac{dv}{dt} = \frac{g}{m} - rv^k \quad \text{for } 0 < k$$

are either intractable or time consuming to solve. What solution properties can we infer (terminal velocity, approach to terminal velocity, etc.)? Some of these answers can be provided analytically, others by simulation. This treatment of the falling body problem also stimulates much discussion and subsequent investigation from my students.

Even when a problem can be treated analytically, answers to specific questions can require the solution of transcendental equations. I mention two examples:

the determination of the time when a falling body (linear model) hits the ground and the determination of the time when a contaminant concentration in a compartmental mixing model achieves a specific value. Subsequent computations may be an important ingredient of the overall problem. Quick answers to these intermediate problems can be provided by technology so that an analysis of the problem solution is then possible. I place the major emphasis on such an analysis instead of on numerical computations or algebraic manipulations.

After first-order problems are studied, numerical techniques for two first-order differential equations,

$$\frac{dx}{dt} = f(t, x, y) \quad \text{and} \quad \frac{dy}{dt} = g(t, x, y)$$

are derived and the associated programs created and saved in the calculator memory. (Again we restrict ourselves to the Euler and improved Euler methods.) Two programs are given to produce graphs of the solutions of initial value problems, one for x (or y) versus t and one for x versus y. The input requirements have the same form as for the first order problem. Again, several graphs may be accumulated in one picture.

One of the key mathematical models studied in this course has the form

$$x'' + B(x') + K(x) = f(t).$$

Such a model arises in the displacement (from equilibrium) motion of a mechanical spring and in electrical flow in an elementary (lumped parameter) circuit. The model generalizes to multiple springs or circuits and is useful in other applications. The usual model assumptions are that the resistive force is linear, i. e., $B(x') = bx'$ with $b \geq 0$, and that the restorative force is linear, i. e., $K(x) = kx$ with $k > 0$. These assumptions make analytical solutions possible and several qualitative properties for the solutions can be studied. For example, for $b > 0$ and $f(t) = 0$, the solutions approach equilibrium $x = x' = 0$ as t increases, and for $b > 0$ and $f(t)$ periodic with period T, the solutions approach a periodic asymptotic state. Such qualitative properties are important in applications. It is quite easy to introduce nonlinear functions for $B(x')$ and $K(x)$ in which the solutions seem to have the same properties. Even though the study of such nonlinear systems is beyond the usual first course in differential equations, students should know to look for such properties. I offer the following alternative models, both with $K(x) = x + .5x^3$ and $f(t) = 5\cos\omega t$,

for simulation studies:

$$B(x') = \begin{cases} .5x', & \text{if } -1 < x' < 1; \\ .25x' + .25, & \text{if } x' \geq 1; \\ .25x' - .25, & \text{if } x' \leq -1 \end{cases}$$

or

$$B(x') = \begin{cases} x' - .5 + (.5 - 1.5x')e^{x'}, & \text{if } x' \leq 0; \\ x' + .5 - (.5 + 1.5x')e^{-x'}, & \text{if } x' > 0. \end{cases}$$

Such models have been shown to possess the desired properties. For the linear case, the periodic output is a cosine function altered by adding a phase angle to get the response $A\cos(\omega t + \Phi)$. (Here A is called the amplitude.) It is interesting to see that additional phase shifts are associated with some, but not all, of these nonlinear models and that resulting amplitude changes also occur. Interesting alternative B, K, and f functions abound. In my classes students are given some library work to check that the observations made in the simulation have been proved.

An introduction to the study of differences between linear and nonlinear problems is also easy using technology. As an example, ask your class to speculate whether the period of motion in an unforced spring (or pendulum) with no damping is a function of the initial conditions. If the spring law is specified, approximations to the period can be obtained by using the integration key on the calculators. The Hooke's law linear model (in contrast to nonlinear models) gives the result that the period does not depend on the initial condition. In fact, by considering several nonlinear springs we can formulate a conjecture concerning which conditions give non-monotonic behavior of these periods. This study can begin in class because a numerical integration program is available for the calculator.

The study of input-output systems can be accomplished primarily in class. Such a study further introduces students in a meaningful context to linear operators. It is easy to generate graphs of the periodic input forcing to a first or second order linear initial value problems with constant coefficients. We then compare the input and output functions. This requires a program to produce appropriate initial values that depend on the value of one or more definite integrals of complicated functions and the subsequent generation of a graph of the output versus t. We consider outputs arising from inputs consisting of piecewise defined functions or of unusual periodic functions such as $\sin(t + \sin(t))$. See the appendix for an example.

In most of my classes I ask students how they will determine the values of unknown parameters in the differential equation models, given that observations are made on the output at various times. For example, if the damping and spring constants in a linear mechanical spring are not known but observations can be made on the solutions at several times, values of the unknown quantities can be made to give a reasonable fit to the data. This problem has been an eye-opener for many of the students. As an introduction to this problem, suppose that a set of observations

$$\{(t_i, x_i) : i = 1, 2, \ldots, n\}$$

is made on the solution which is known to have the form

$$x = ae^{-pt} + be^{-qt}.$$

Here a, b, p, and q are unknown parameters. If data is given so that the asymptotic approach to 0 can be seen, it is possible to get a good idea of the values of the parameters by informal methods. Then a presentation is made to show we can refine these values in a least squares fit to the data which will require us to solve several nonlinear equations in the unknown parameters. Newton's method (in several variables) is an extension of a familiar algorithm and can be easily programmed on the calculator. If the spring model is nonlinear, the differential equation cannot be solved exactly and even more interesting aspects arise and can be handled using technology (see [2]). These types of exercises have extended the abilities of many students.

A topic now accessible using the calculators is the chaotic solution of discrete dynamical systems or chaotic solutions of differential equations in three or more dependent variables. The first case presented is the asymptotic or steady state solutions of a first order system given by

$$y_{n+1} = (1 + a)y_n - ay_n^2, \quad y_0 = 0.1.$$

For fixed values of the parameter a between 1.8 to 3, the solutions have period one, period two, period four, period eight, etc, finally cycling through a "Cantor" set. Julia sets are "graphed" for discrete systems in the complex plane. Finally, Lorentz's weather model is introduced and an apparently nonrepeating solution is calculated and graphed. This work is followed by a presentation of the NOVA film on chaotic solutions of various problems. In this way students are introduced to a timely modern topic which is described in popular journals.

5 Linear Systems of Differential Equations

One of the problems in a standard course in differential equations is to find the general solution of a system $dx/dt = Ax$. In most courses the problem is presented with heavy emphasis on the case in which x has two components because the calculations for the solution are involved and time-consuming. The calculator is a tool that removes this computational burden and permits us to study systems with three or more components. This addition allows me to indicate peculiarities which may arise in these cases.

Algorithms for the problem $dx/dt = Ax$ that are developed in our course are:

1. A solution technique for a system $Ax = b$. If A is nonsingular this can be handled with keystroke commands. Such a procedure is first encountered by students in many environments, some of which are not mathematics classrooms. Appropriate application of the process is not guaranteed. We illustrate the problems that may arise, but a careful presentation is postponed to another course. A systematic and interactive method for manipulating the equations $Ax = b$ (Gauss-Jordan elimination) is used to assure that solubility is detected at appropriate stages of the problem. In particular, the procedure is used to find nontrivial solutions of a system $Ax = 0$.

2. A partial algorithm to find the eigenvalues and corresponding eigenvectors of a matrix A. The eigenvalue equation is displayed, a combination of graphical methods and the calculator's SOLVE algorithm is used to solve the eigenvalue equation, and the vector equation $(A - rI)c = 0$ is solved for a corresponding eigenvector c.

3. The general solution of $dx/dt = Ax$,

$$x = \sum_{i=1}^{n} \alpha_i e^{r_i t} c_i$$

is constructed. Here the c_i quantities are the eigenvectors and the α_i are constants to be specified by initial conditions.

Each of the above procedures is computationally intensive. When done by hand, mistakes often prevent the students from proceeding to analyze the answer, which is considered a very important ingredient of the problem. For example, when $n = 2$, the components of x can be shown in the phase plane. Students should be able to predict the behavior for simple situations. If the eigenvalues are negative real numbers $r_1 < r_2 < 0$, then the solution x will proceed from the initial configuration to asymptotic behavior along the vector c_2. Other examples that are studied include behavior in the neighborhood of saddle points or spiral points.

We also study the non-homogeneous system $dx/dt = Ax + f(t)$, particularly for the case when $f(t)$ is a combination of simple elementary functions and possibly is periodic. In some cases, the method of undetermined coefficients leads to the solution of matrix-vector equations, an easy task at this point. In the case of periodic input with somewhat complicated $f(t)$, the variation of constants procedure can be used to calculate the initial conditions that give periodic output.

Some nonlinear systems $dw/dt = F(w)$ are studied. Critical points are located, sometimes with the use of Newton's method if a good starting point can be found. Symbolic capabilities are especially important in creating the derivative ingredients of Newton's method, in evaluating functions at successive trial points, and in repeatedly finding the solutions of the associated linear systems. Classification of critical points as to type (node, saddle, spiral, etc.) are made by studying linearizations about the critical point. Here again, accuracy in calculating derivatives and in function evaluations is crucial. Simulations in the phase plane emphasize the use of these linearizations. Stable isolated periodic solutions are demonstrated for simple systems. The ability to do the necessary numerical and algebraic calculations rapidly enables students to begin a qualitative analysis of these systems and would be impossible to do without these tools.

6   Conclusions

Factors that set the calculator-enhanced course apart from the usual class presentation include: (a) class participation, (b) greater emphasis on topics from previous mathematics courses such as numerical integration, arc length, evaluation of Taylor series for functions other than the few that can be done reliably by hand, etc., (c) more interesting application problems, and (d) much more inspection of the answers obtained. The objectives of the present enhancement

include: increased student motivation, an increased retention of the topics that are studied, and an improvement in the student's perspective of the scope of calculus.

The use of technology and student projects enable us to attack problems that require several concepts. The individual techniques (for example, the construction of graphs, the calculation of a Jacobian, or a numerical integration) are often known to the students, but the amount of work prevents the study of interesting applications of the material at hand unless technology is used. Problems that require complex number and matrix calculations or graphical constructions are quite easy when the calculator is used.

I should note that an emphasis on such problems and the subsequent analysis is not always popular with students. Many are content to obtain a formula answer for routine problems and do not use their knowledge of functions to check the work. Early class work contains emphasis on the construction of graphs of one or more of the solutions of initial value problems. Much of the heavier problem assignment is contained in the projects scheduled at the 1/3 and 2/3 semester marks.

The calculator-enhanced course has been offered six times to a total of 150 students and we have received evaluations from 88% of these students. One of the classes was an honors class, one was a summer class which included several students who were repeating the course, and the others were regular classes. Sample responses on the student evaluation are as follows:

1. Statement: The graphics calculator helped me understand the material in the course. 70% of the students agreed, 15% disagreed.

2. Statement: The graphics calculator allowed me to do more exploration and investigation in solving problems. 65% of the students agreed, 15% disagreed.

3. Statement: Learning the graphics calculator was so difficult that it detracted from learning the material in the course. 15% of the students agreed, 70% disagreed.

4. Statement: I could have learned more if I had not used the graphics calculator. 15% of the students agreed, 70% disagreed.

5. Statement: I would recommend that entering freshman seek out courses using the graphics cal-

culator. 60% of the students agreed, 15% disagreed.

References

1. LaTorre, D. R., *et al.* (1990). "Calculator-Based Calculus," in *Priming the Calculus Pump: Innovations and Resources*, Tucker, T. (Ed.), MAA Notes Number 17, Mathematical Association of America, Washington, D. C.

2. Proctor, T. G. (1992). *Calculator Enhancement for Differential Equations, A Manual of Applications using the HP-48S and HP-28S Calculators*, Saunders College Publishing, Philadelphia.

Appendix: Some Problems

Problem 1. The following differential equations have served as possible models for the growth of a population: the logistic model,

$$\frac{dp}{dt} = -rp(1 - p/K),$$

the Gompertz model

$$\frac{dp}{dt} = rp\ln(K/p),$$

and a model constructed for a particular inflection position,

$$\frac{dp}{dt} = \text{IFTE}(p \le .75 * K, (.75 * K)^2 - (p - .75 * K)^2,$$
$$(.75 * K)^2(1 - 16(p - .75 * K)^2/K^2)).$$

Now consider the following model

$$\frac{dp}{dt} = .5p(e^{\sin(t)} - p).$$

Set the plotting screen to show

$$-.5 \le t \le 12.56, \quad -.5 \le p \le 3$$

and plot the solution that has initial condition $p(0) = 1$. Use that solution to guide you in making several other choices for $p(0)$. Can you find an attracting periodic solution? Can you think of this differential equation as modelling a population growth if we change the scale in p and t? What feature of a population model does the $e^{\sin(t)}$ term model? What is the significance of the periodic solution?

Problem 2. (I have chosen to show this problem for second order differential equations, but there are first order and vector versions. Of course the parameter values and input function $f(t)$ mentioned below are prescribed.) Recall that the solution of the nonhomogeneous second order differential equation:

$$\frac{d^2x}{dt^2} + 2b\frac{dx}{dt} + \omega^2 x = f(t),$$

$$x(0) = \frac{dx}{dt}(0) = 0, \ \omega^2 > b^2.$$

is given by $x_p(t) =$

$$\frac{1}{\sqrt{\omega^2 - b^2}} \int_0^t e^{-b(t-s)} \sin\sqrt{\omega^2 - b^2}(t - s)f(s)ds.$$

Note that our prescribed forcing function $f(t)$ is periodic with period length T. Rechoose the initial conditions so that $x(T) = x(0)$ and $x'(T) = x'(0)$ and graph the solution over one period. (The students should note the resulting solution is periodic and since the damping coefficient $b > 0$, all solutions will eventually be a close approximate of the periodic solution when viewed over one period.)

Here is part of what is required for the solution. The functions

$$x_1(t) = e^{-bt}[\cos \mu t + (b/\mu) \sin \mu t]$$

$$x_2(t) = (1/\mu)e^{-bt} \sin \mu t$$

are solutions of the associated homogeneous system and

$$x_1(0) = x_2'(0) = 1$$

and

$$x_1'(0) = x_2(0) = 0,$$

and a general solution is

$$x(t) = ux_1(t) + vx_2(t) + x_p(t)$$

where $x_p(t)$ is the solution given above. The calculator is used to compute $x_p(T)$ and $x_p'(T)$ numerically and to solve the periodicity condition for a and b:

$$\begin{bmatrix} 1 - x_1(T) & -x_2(T) \\ -x_1'(T) & 1 - x_2'(T) \end{bmatrix} \begin{bmatrix} a \\ b \end{bmatrix} = \begin{bmatrix} x_p(T) \\ x_p'(T) \end{bmatrix}$$

and the graph is constructed.

Problem 3. Suppose that $x(t)$ satisfies the initial value problem

$$\frac{d^2x}{dt^2} + f(x) = 0, \quad x(0) = z, \quad \frac{dx}{dt}(0) = 0.$$

where the essential feature of $f(x)$ is that it changes sign from negative to positive as x increases through zero.

1. Show that

$$\frac{dx}{dt} = \pm\sqrt{F(z) - F(x)},$$

    where

$$F(x) = 2\int_0^x f(s)\,ds.$$

    If we denote by $T/2$ the time for the trajectory to proceed from the starting point to the state $x(T/2) = z_1$, $\frac{dx}{dt}(T/2) = 0$, then

$$T = 2\int_0^{T/2} dt = 2\int_{z_1}^z \frac{dx}{\sqrt{F(z) - F(x)}}.$$

2. Find the periods T for the functions $f(x)$ and the starting points indicated:

    (a) $f(x) = x + x^3$: $F(x) = x^2 + x^4/2$, and $z_1 = -z$, $z = 1, 2, 3$

    (b) $f(x) = \sin x$: $F(x) = 2[1 - \cos x]$, and $z_1 = -z$, $z = 1, 2, 3$.

Problem 4. First the following material must be given to the students: Suppose the data

$$\{(t_i, y_i) : i = 1, 2, \ldots, N\}$$

is given for $y = g(t, p, q)$, (e. g., the solution of an initial value problem) and we wish to choose the values of p and q to obtain the best fit to the data, say in a least squares sense. That is, we want to find p and q to minimize the sum

$$\sum_{i=1}^N [y_i - g(t_i, p, q)]^2.$$

By taking derivatives with respect to p and q and setting them to 0 we obtain:

$$\sum_{i=1}^N [y_i - g(t_i, p, q)]\frac{\partial g}{\partial p}(t_i, p, q) = 0$$

and

$$\sum_{i=1}^N [y_i - g(t_i, p, q)]\frac{\partial g}{\partial q}(t_i, p, q) = 0.$$

You should view this as a special case of the problem of finding a solution of a vector equation $F(w) =$

0 of a vector variable w. (In this application we have 2 parameters, w will be the 2 vector of parameters and F will be the 2 vector of partial derivatives.) If we have an approximate solution w_0, then Taylor's theorem gives the approximate formula

$$F(w) = F(w_0) + J(w_0)(w - w_0)$$

where the matrix J has i, j element $\frac{\partial F_i}{\partial w_j}$. If w is to be a good approximation of the solution, the left side of this equation is zero and we get a "formula" for an improved vector solution w in terms of the old approximate w_0.

At this point a specific form for g, for example

$$g(t, p, q) = e^{-pt} - 2e^{-qt},$$

and a data set of observations is given. The students are expected to create calculator programs for

$$[y - g(t, p, q)]\frac{\partial g}{\partial p}(t, p, q)$$

and

$$[y - g(t, p, q)]\frac{\partial g}{\partial q}(t, p, q).$$

Following this, the student uses a program called DER for finding the derivatives of these functions with respect to p and q and forming terms in the Hessian matrix J. Next the list

$$\{(t_1, y_1), (t_2, y_2), \ldots, (t_N, y_N)\}$$

is stored as DTA1. After assigning preliminary values to p and q, programs called JACM and FACM are executed to accumulate the data sums in the Hessian matrix and the function $F(w)$. Now the system

$$\begin{bmatrix} p_{\text{new}} \\ q_{\text{new}} \end{bmatrix} = \begin{bmatrix} p \\ q \end{bmatrix} - J(p, q)^{-1}F(p, q)$$

is solved for new values of p and q. The last step is repeated until the values of p and q are satisfactory.

Part IV:

Symbolic Computation in Advanced Undergraduate Courses

Teaching Combinatorics and Graph Theory with Mathematica

Steven S. Skiena

Department of Computer Science

State University of New York, Stony Brook

The rush that comes from discovering something new is what makes learning exciting. The pedagogical ideal is to provide a learning environment where students are encouraged to make their own discoveries. I believe that the recent availability of computer tools for mathematics, such as symbolic mathematics systems, will have their most profound impact by making it easy for students to experiment with and explore mathematics, the way chemistry sets and toy microscopes encourage children to explore science.

Since theorems in discrete mathematics often start out as observations, I cannot imagine branches of mathematics that are better suited to experimentation than combinatorics and graph theory. To facilitate such exploration, we have developed *Combinatorica* [17], a computational environment for combinatorics and graph theory. Designed for both teaching and research use, and running under *Mathematica*, *Combinatorica* provides a new way to get students excited about discrete mathematics. It can turn students into researchers, by giving them the power to ask questions about particular structures and quickly get answers. Properly motivated, students can be lead to discover and prove theorems that they can call their own.

In the rest of this paper, we describe *Combinatorica* in more detail, as well as other systems for discrete mathematics. We then proceed to show, through examples, how some of the properties of permutations can be made to come alive with *Combinatorica*. We conclude by presenting a variety of other interesting problems which can be approached via *Combinatorica* in a non-traditional way.

1 Combinatorica

Combinatorica, a workbench for manipulating graphs and other combinatorial objects has, since its initial release in May 1990, already become perhaps the most widely used combinatorial computing environment. There are 5500 copies of the associated book [17] in print, and over 250 subscribers to the associated electronic mailing list. *Combinatorica* was recently recognized with a 1991 EDUCOM Higher Education Software Award, for distinguished mathematics software. Users include students, teachers, and researchers from a variety of fields including mathematics, computer science, engineering, and the natural sciences.

Combinatorica is built on top of *Mathematica*, comprising over 230 combinatoric and graph theoretic functions. Specifically, it provides functions for constructing a variety of combinatorial objects, determining their properties, and displaying them in interesting ways. All functions are accessible by the command-line interface of *Mathematica*, and can also be called as subroutines by user-written programs.

Combinatorica requires *Mathematica*, version 1.2 or later, and the latest release is available via anonymous ftp from cs.sunysb.edu. Beginning with version 2.0, *Combinatorica* is included in the standard distribution of *Mathematica*, in the Packages/DiscreteMath directory, so no special effort is necessary to acquire it. For those running earlier versions of *Mathematica* and without network access, Macintosh and MS-DOS disks containing the software are available for $15.00 from Wolfram Research, Inc. To be placed on the electronic mailing list, please send a request to skiena@sbcs.sunysb.edu.

AcyclicQ	AddEdge	AddVertex	AllPairsShortestPath
ArticulationVertices	Automorphisms	Backtrack	BiconnectedQ
BinarySearch	BinarySubsets	BipartiteMatching	BipartiteQ
BreadthFirstTraversal	Bridges	CartesianProduct	CatalanNumber
ChangeEdges	ChangeVertices	ChromaticNumber	ChromaticPolynomial
CirculantGraph	CircularVertices	CliqueQ	CodeToLabeledTree
Cofactor	CompleteQ	Compositions	ConnectedComponents
ConnectedQ	ConstructTableau	Contract	Cycle
DeBruijnSequence	DegreeSequence	DeleteCycle	DeleteEdge
DeleteFromTableau	DeleteVertex	DepthFirstTraversal	DerangementQ
Derangements	Diameter	DilateVertices	DistinctPermutations
Distribution	DurfeeSquare	Eccentricity	EdgeColoring
EdgeConnectivity	Edges	EmptyGraph	EmptyQ
EncroachingListSet	EquivalenceClasses	EquivalenceRelationQ	EulerianCycle
EulerianQ	Eulerian	ExactRandomGraph	ExpandGraph
FerrersDiagram	FindCycle	FromAdjacencyLists	FromCycles
FromInversionVector	FromOrderedPairs	FromUnorderedPairs	FunctionalGraph
Girth	GraphCenter	GraphComplement	GraphDifference
GraphIntersection	GraphJoin	GraphPower	GraphProduct
GraphSum	GraphUnion	GraphicQ	GrayCode
GridGraph	HamiltonianCycle	HamiltonianQ	Harary
HasseDiagram	HeapSort	Heapify	HideCycles
Hypercube	IdenticalQ	IncidenceMatrix	IndependentSetQ
Index	InduceSubgraph	InsertIntoTableau	IntervalGraph
InversePermutation	Inversions	InvolutionQ	IsomorphicQ
Isomorphism	Josephus	KSubsets	K
LabeledTreeToCode	LexicographicPermutations	LexicographicSubsets	LineGraph
LongestIncreasingSubsequence	M	MakeGraph	MakeSimple
MakeUndirected	MaximalMatching	MaximumAntichain	MaximumClique
MaximumIndependentSet	MaximumSpanningTree	MinimumChainPartition	MinimumChangePermutations
MinimumSpanningTree	MinimumVertexCover	MultiplicationTable	NetworkFlowEdges
NetworkFlow	NextComposition	NextKSubset	NextPartition
NextSubset	NextTableau	NormalizeVertices	NthPair
NthPermutation	NthSubset	NumberOfCompositions	NumberOfDerangements
NumberOfInvolutions	NumberOfPartitions	NumberOfPermutationsByCycles	NumberOfSpanningTrees
NumberOfTableaux	OrientGraph	PartialOrderQ	PartitionQ
Partitions	Path	PerfectQ	PermutationGroupQ
PermutationQ	Permute	PlanarQ	Polya
PseudographQ	RadialEmbedding	Radius	RandomComposition
RandomGraph	RandomHeap	RandomKSubset	RandomPartition
RandomPermutation	RandomSubset	RandomTableau	RandomTree
RandomVertices	RankGraph	RankPermutation	RankSubset
RankedEmbedding	ReadGraph	RealizeDegreeSequence	RegularGraph
RegularQ	RemoveSelfLoops	RevealCycles	RootedEmbedding
RotateVertices	Runs	SamenessRelation	SelectionSort
SelfComplementaryQ	ShakeGraph	ShortestPathSpanningTree	ShortestPath
ShowGraph	ShowLabeledGraph	SignaturePermutation	SimpleQ
Spectrum	SpringEmbedding	StableMarriage	Star
StirlingFirst	StirlingSecond	Strings	StronglyConnectedComponents
Subsets	TableauClasses	TableauQ	TableauxToPermutation
Tableaux	ToAdjacencyLists	ToCycles	ToInversionVector
ToOrderedPairs	ToUnorderedPairs	TopologicalSort	TransitiveClosure
TransitiveQ	TransitiveReduction	TranslateVertices	TransposePartition
TransposeTableau	TravelingSalesmanBounds	TravelingSalesman	TreeQ
TriangleInequalityQ	Turan	TwoColoring	UndirectedQ
UnweightedQ	V	VertexColoring	VertexConnectivity
VertexCoverQ	Vertices	WeaklyConnectedComponents	Wheel
WriteGraph			

Figure 1: Functions in *Combinatorica*.

Combinatorica includes functions for constructing such combinatorial objects as permutations, subsets, partitions, and Young tableaux. These objects can be constructed both canonically and randomly, with functions to rank, unrank, and generate the successor to each object. Important properties of these objects, such as the number of inversions or the longest increasing subsequence of a permutation, can be computed. Further, bijections between different objects have been implemented. For example, the unique permutation associated with two Young tableaux of the same shape can be readily constructed and displayed.

The graph theoretic component of *Combinatorica* includes functions for constructing a wide variety of graphs, such as stars, cycles, wheels, trees, hypercubes, circulant and random graphs. Further, there are a significant collection of operations which construct graphs from other graphs, including line graphs, graph products, joins, and powers. Most interesting graphs can thus be easily specified. Predicates test whether the graph is acyclic, biconnected, bipartite, Eulerian, Hamiltonian, planar, or have any of a variety of other characteristics. Invariants of graphs such as chromatic number, connectivity, diameter, and radius can be computed, along with such algorithmic properties as minimum spanning tree, maximum flow, maximum matching, and shortest path. Finally, each graph can be displayed in several different ways to determine the most appropriate representation, including circular, ranked, radial, random, and spring embeddings. A complete list of *Combinatorica* functions appears in Figure 1.

To illustrate the way the combinatorial and graph theoretic functions of *Combinatorica* interact with the underlying *Mathematica* programming language, the Boolean lattice in Figure 2 was constructed with the command:

```
HasseDiagram[
    MakeGraph[
        Subsets[6],
        ((Intersection[#2,#1] === #1)
            && (#1 != #2))&
    ]
]
```

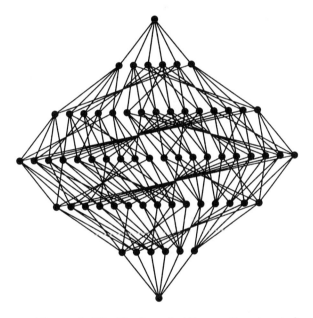

Figure 2: The Boolean Lattice as Constructed with *Combinatorica*.

To explain, Subsets generates a list of all subsets of n elements, which MakeGraph uses as vertices of a graph. Edges are defined between any two subsets s_1 and s_2, where $s_1 \subset s_2$. This binary relation is specified as a pure function in the underlying programming language. Associated with each graph is an embedding, and HasseDiagram changes the default embedding to that shown in Figure 2.

Combinatorica is implemented in *Mathematica* [20], and comprises over 230 functions representing about 2500 lines of code. The last statistic is quite revealing—through careful programming and the fact that *Mathematica* is a high-level language, the average documented function is only eleven lines long. *Combinatorica* is the result of two years of development, primarily by the author, with the assistance of Anil Bhansali over part of this period.

2 Software for Combinatorial Computing

The idea of a system for experimenting with combinatorial mathematics is natural enough that many such tools have been built over the years. In this section, we will survey some of these systems that might be of interest to educators. Be forewarned that most of the systems have not been widely distributed, and that the quality and appropriateness for educational use varies considerably. All of these programs should be either in the public domain or available at a modest fee. Some other packages, in BASIC, for teaching discrete mathematics are described in [16]. None of the systems discussed here, except *Combinatorica*, significantly support combinatorial structures other than graphs.

We are aware of three textbooks in discrete mathematics that provide significant computer support. *Implementing Discrete Mathematics* [17], describes *Combinatorica*. Also appropriate is Stanton and White's *Constructive Combinatorics* [19], which provides Pascal programs for a variety of combinatorial functions. The third book, *Learning Discrete Mathematics with ISETL* [1] focuses on sets and functions, with limited coverage of combinatorics and graph theory.

We distinguish between two types of graph theoretic software: interactive systems and libraries. Interactive systems are characterized by graph editors, which permit the user to literally draw a graph on the screen using a mouse, to add/delete vertices and edges. Most attempts to produce interactive tools for discrete mathematics have focused on developing graph editors. *CABRI* [6] is a French program for the Apple Macintosh with a very slick graph editor and a nice but small collection of invariants. *GraphLab* is an extensible graph editor for the NeXT [4], which has been undergoing continued development. *SPREMB* [7], developed by researchers at the University of Queensland in Australia, comprises a graph editor *GED* for Sun workstations and a collection of C language filter programs for operations on embeddings of graphs. *Graph^{Ed}* [11], from Universität Passau, is another SunTools-based graph editor designed for manipulating graph grammars and entity-relationship diagrams. We note that *Combinatorica* currently does not include a graph editor, and instead reads and writes files in *SPREMB* format, so we can use *GED* to edit graphs.

Several systems, in addition to *Combinatorica*, provide additional facilities. *GRAPPLE* is a system of Pascal programs for manipulating graphs for IBM PC-compatible computers, built by students at Indiana University/Purdue University at Fort Wayne under the direction of Prof. Marc Lipman. *Nauty* [13] is a set of C language procedures for determining the automorphism group of a graph. *NETPAD*, under development at Bellcore [5], provides a graphical, menu-driven interface for manipulating and analyzing graphs, and has been used educationally at both the high school and college levels. *GraphPack* [12] contains a library of graph algorithms, written in an extension of C called *LiLa*, and includes an integral graph editor and interpreter. Both *NETPAD* and *GraphPack* require a workstation and windowing system to run.

Finally, there are several efforts that don't fit easily into any category. There are several books that provide implementations of a significant number of combinatorial algorithms in a variety of programming languages, including Pascal [5], C [10], and FORTRAN [14]. Algorithm animation uses computer graphics to illustrate the behavior of combinatorial algorithms and data structures, with *BALSA* [2, 3] and *TANGO* [18] being representative algorithm animation systems. An especially interesting program is *GRAFFITI* [8, 9], which generalizes from a collection of graphs to form its own conjectures. Fortunately, mathematicians are still necessary to prove the theorems and write papers, several of which have resulted from *GRAFFITI* conjectures. With the addition of appropriate embeddings, we have converted all 195 graphs of *GRAFFITI* to our format, providing an extensive library of graphs for testing conjectures with *Combinatorica*.

3   Permutations with Combinatorica

In this section, we will illustrate how instruction in discrete mathematics can be enhanced using a package like *Combinatorica*. We will focus on one particular topic, permutations, primarily because it demonstrates the versatility of what is largely a graph theoretic tool. The first chapter of [17] discusses permutations and subsets, and is the source of most of the material in this section.

In the examples that follow, we provide a series of *Mathematica/Combinatorica* commands and the associated output. While the commands may appear cryptic to someone unfamiliar with *Mathematica*, in fact they are simple to generate once you are familiar with *Mathematica* function definitions and iteration primitives like Map.

3.1   Permutations from Transpositions

Perhaps the most fundamental combinatorial object is the **permutation**. We define a permutation as an ordering of the integers 1 to n without repetition. A permutation of items defines a particular arrangement of them, and so permutations can be viewed as either objects (the actual arrangement) or operations (the rearrangement necessary to move from one arrangement to another). Permutations thus provide an interface between algebra and combinatorics.

There are many ways to generate all permutations of n items. One of the most interesting is to sequence permutations to minimize the amount of change between adjacent permutations. Since two distinct permutations contain the same elements in a different order, the minimum change between two distinct permutations consists of an exchange of two elements. In fact, it is always possible to arrange the permutations such that every two adjacent permutations differ by exactly one exchange or **transposition**.

To best illustrate the transposition structure of permutations, we will introduce some of the graph theoretic tools of *Combinatorica*. A **graph** is a collection of pairs of elements from some set. If the elements of the set are represented by points or **vertices** and the pairs by lines or **edges** connecting the points, we can get a visual representation of the underlying structure. We may construct a graph where the vertices correspond to permutations, with an edge between permutations that differ by exactly one transposition. The following *Combinatorica* interactions illustrate how some properties of permutations can be investigated.

1. The transposition graph illustrates that each permutation is one transposition removed from $\binom{n}{2}$ other permutations. Further, tracing all the paths in the graph between two vertices shows that any particular pair of permutations differ from each other by either an even or odd number of transpositions, independent of the order in which the transpositions are applied.

```
In[1]:= ShowGraph[ r = RankedEmbedding[ MakeGraph[
   Permutations[{1,2,3,4}], (Count[#1-#2,0]==(Length[
   #1]-2))& ], {1} ] ];
```

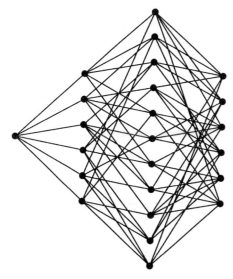

2. All cycles in the transposition graph are of even length, meaning that the graph is **bipartite**. This accounts for the even-odd transposition property.

```
In[2]:= BipartiteQ[r]
Out[2] = True
```

3. A **Hamiltonian cycle** of a graph is a tour that visits each vertex of the graph without repetition. The Hamiltonian cycle of a transposition graph corresponds to an ordering where each successive permutation differs by exactly one transposition. Since a minimum amount of work has to be done to construct each subsequent permutation from its predecessor, minimum change order is the most efficient way to construct permutations.

```
In[3]:= Permutations[{1,2,3,4}] [[ HamiltonianCycle[r] ]]
Out[3]= {{1, 2, 3, 4}, {1, 2, 4, 3}, {1, 3, 4, 2},
   {1, 3, 2, 4}, {1, 4, 2, 3}, {1, 4, 3, 2}, {2, 4, 3, 1},
   {2, 1, 3, 4}, {2, 1, 4, 3}, {2, 3, 4, 1}, {2, 3, 1, 4},
   {2, 4, 1, 3}, {3, 4, 1, 2}, {3, 1, 4, 2}, {3, 1, 2, 4},
   {3, 2, 1, 4}, {3, 2, 4, 1}, {3, 4, 2, 1}, {4, 3, 2, 1},
   {4, 1, 2, 3}, {4, 1, 3, 2}, {4, 3, 1, 2}, {4, 2, 1, 3},
   {4, 2, 3, 1}, {1, 2, 3, 4}}
```

4. The permutations of n elements, for $n \geq 4$, can be sequenced in **maximum change order** where neighboring permutations differ in all positions. Such a sequence can be constructed by finding a Hamiltonian cycle on the appropriate adjacency graph. This graph is disconnected for $n = 3$.

```
In[4]:= Permutations[{1,2,3,4}] [[ HamiltonianCycle[
   MakeGraph[Permutations[{1,2,3,4}], (Count[#1-#2,0] ==
   0)&] ] ]]
Out[4]= {{1, 2, 3, 4}, {2, 1, 4, 3}, {1, 3, 2, 4},
   {2, 4, 1, 3}, {1, 3, 4, 2}, {2, 1, 3, 4}, {1, 2, 4, 3},
   {2, 3, 1, 4}, {1, 4, 2, 3}, {2, 3, 4, 1}, {1, 4, 3, 2},
   {3, 1, 2, 4}, {2, 4, 3, 1}, {4, 1, 2, 3}, {3, 2, 1, 4},
   {4, 1, 3, 2}, {3, 2, 4, 1}, {4, 3, 1, 2}, {3, 4, 2, 1},
   {4, 2, 1, 3}, {3, 1, 4, 2}, {4, 2, 3, 1}, {3, 4, 1, 2},
   {4, 3, 2, 1}, {1, 2, 3, 4}}
```

5. The permutations of this multiset are returned in lexicographic order, so it is apparent that each is constructed only once. There are $\binom{6}{3}$ permutations of this multiset, since there are that many ways to select the position of the ones.

```
In[5]:= DistinctPermutations[{1,1,1,2,2,2}]
Out[5]= {{1, 1, 1, 2, 2, 2}, {1, 1, 2, 1, 2, 2},
   {1, 1, 2, 2, 1, 2}, {1, 1, 2, 2, 2, 1},
   {1, 2, 1, 1, 2, 2}, {1, 2, 1, 2, 1, 2},
   {1, 2, 1, 2, 2, 1}, {1, 2, 2, 1, 1, 2},
   {1, 2, 2, 1, 2, 1}, {1, 2, 2, 2, 1, 1},
   {2, 1, 1, 1, 2, 2}, {2, 1, 1, 2, 1, 2},
   {2, 1, 1, 2, 2, 1}, {2, 1, 2, 1, 1, 2},
   {2, 1, 2, 1, 2, 1}, {2, 1, 2, 2, 1, 1},
   {2, 2, 1, 1, 1, 2}, {2, 2, 1, 1, 2, 1},
   {2, 2, 1, 2, 1, 1}, {2, 2, 2, 1, 1, 1}}
```

6. A built-in *Mathematica* function computes multinomial coefficients and can be used to determine the number of permutations of a multiset.

```
In[6]:= Multinomial[3,3]
Out[6]= 20
```

This kind of experimentation suggests several exercises for students, appropriate for the junior/senior level. Can you prove that all cycles in the transposition graph are of even length? What do the minimum and maximum difference permutation graphs look like for $n = 3$? Which permutation requires the most transpositions to get to the identity permutation, and how many are required? What is the general form of the multinomial coefficient when you have more than two different classes of objects?

3.2 Permutation Groups

A mathematical **group** is a set of objects that is closed under an associative multiplication operator, that contains an identity element, and for which every element has an inverse. The complete set of permutations defines a group, and other permutation groups are defined by appropriate subsets of permutations.

For binary operators over a closed set of elements, it is useful to define a **multiplication table** showing how every pair of elements relates to each other. For permutation groups, Permute serves as the multiplication operator, and the permutation Range[n] as the identity element.

1. The complete set of permutations on n elements, or **symmetric group** S_n, is closed under composition. The multiplication table shows that the first permutation is the identity element, each permutation has an inverse, and that the set of permutations is closed under multiplication. Since the table is not symmetric around the diagonal, S_n is not a commutative or **abelian** group.

```
In[1]:= TableForm[ MultiplicationTable[
  Permutations[{1,2,3}], Permute ] ]
Out[1]//TableForm=  1  2  3  4  5  6
                    2  1  5  6  3  4
                    3  4  1  2  6  5
                    4  3  6  5  1  2
                    5  6  2  1  4  3
                    6  5  4  3  2  1
```

2. Star[n] is the graph of $n-1$ vertices of degree 1 and 1 vertex of degree $n-1$.

```
In[2]:= ShowLabeledGraph[ Star[5] ];
```

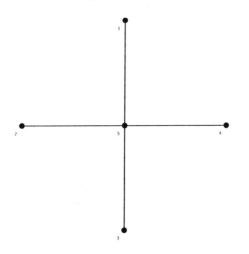

3. An **automorphism** is a permutation of the vertices so that the graph is isomorphic to itself. The automorphisms of a graph always describe a group.

```
In[3]:= star = Automorphisms[Star[5]]
General::spell1:
Out[3]={{1, 2, 3, 4, 5}, {1, 2, 4, 3, 5},
  {1, 3, 2, 4, 5}, {1, 3, 4, 2, 5}, {1, 4, 2, 3, 5},
  {1, 4, 3, 2, 5}, {2, 1, 3, 4, 5}, {2, 1, 4, 3, 5},
  {2, 3, 1, 4, 5}, {2, 3, 4, 1, 5}, {2, 4, 1, 3, 5},
  {2, 4, 3, 1, 5}, {3, 1, 2, 4, 5}, {3, 1, 4, 2, 5},
  {3, 2, 1, 4, 5}, {3, 2, 4, 1, 5}, {3, 4, 1, 2, 5},
  {3, 4, 2, 1, 5}, {4, 1, 2, 3, 5}, {4, 1, 3, 2, 5},
  {4, 2, 1, 3, 5}, {4, 2, 3, 1, 5}, {4, 3, 1, 2, 5},
  {4, 3, 2, 1, 5}}
```

4. The two equivalence classes correspond to the vertices of different degree. The 24 automorphisms listed above represent all the permutations of the four degree 1 vertices.

```
In[4]:= EquivalenceClasses[ SamenessRelation[star] ]
Out[4]={{1, 2, 3, 4}, {5}}
```

5. Not all subsets of the symmetric group are permutation groups. Two permutations form a group only if one is the identity element and the other is an **involution**, a permutation that is its own inverse.

```
In[5]:= PermutationGroupQ[{{1,2,3,4},{4,2,3,1}}]
Out[5]= True
```

More lines of investigation now suggest themselves. What does the automorphism group of other simple graphs look like, like cycles, wheels, and paths? Can you prove that the identity permutation and an involution form a group? Which subsets of S_3 form permutation groups?

3.3   Cycle Structure

Suppose multiplying by a particular permutation takes element p_i to p_j and p_j to p_k. Eventually, we always get back to the first element p_i, forming a cycle. Any permutation can be considered as a composition of disjoint cycles, where each cycle specifies a circular shift of the elements to be permuted. For example, the permutation $p = \{5, 3, 4, 2, 6, 1\}$ has a cycle structure of $\{\{5, 6, 1\}, \{3, 4, 2\}\}$. Observe that multiplying a permutation by p takes the first element to the fifth position, the fifth element to the sixth position, and the sixth element to the first position, closing the first cycle. This suggests an alternative representation for permutations in terms of their **cycle structure**.

1. The identity permutation consists of n singleton cycles or fixed-points.

```
In[1]:= ToCycles[{1,2,3,4,5,6,7,8,9,10}]
Out[1]= {{1}, {2}, {3}, {4}, {5}, {6},
  {7}, {8}, {9}, {10}}
```

2. The reverse of the identity permutation contains only cycles of length two, with one singleton cycle if n is odd. Cycles of length two are transpositions.

```
In[2]:= ToCycles[ Reverse[Range[10]] ]
Out[2]= {{10, 1}, {9, 2}, {8, 3}, {7, 4}, {6, 5}}
```

3. Any permutation p with a maximum cycle length of two is an involution, meaning that multiplying p by itself gives the identity.

```
In[3]:= Permute[ Reverse[Range[10]], Reverse[Range[10]] ]
Out[3]= {1, 2, 3, 4, 5, 6, 7, 8, 9, 10}
```

4. Since each cycle structure defines a unique permutation, converting to and from the cycle structure is an identity operation.

```
In[4]:= Apply[And, Table[ p=RandomPermutation[20];
  p===FromCycles[ToCycles[p]], {50}] ]
Out[4]= True
```

Can you prove that the involutions are exactly the permutations of maximum cycle length two? Which permutations consist of a single cycle, and how many of them are there? How many cycles does a random permutation usually have? Can you find a recurrence relation to count the number of permutations of length n with exactly k cycles?

3.4 Signatures

As discussed earlier, permutations can be constructed as a sequence of transpositions, or swaps of two elements, starting from the identity permutation. The parity of the size of any such sequence of transpositions is invariant for any particular permutation. The **signature** or **sign** of a permutation p is $+1$ if the number of transpositions in the sequence is even and -1 if it is odd.

1. One transposition is sufficient to move this to the identity permutation, meaning an odd number of transpositions no matter how they are selected.

```
In[1]:= SignaturePermutation[{1,3,2,4,5,6,7,8}]
Out[1]= -1
```

2. All permutations have the same sign as their inverse.

```
In[2]:={SignaturePermutation[p=RandomPermutation[50]],
  SignaturePermutation[InversePermutation[p]]}
Out[2]= {1, 1}
```

3. The set of all even permutations forms a group. This proves that the product of two even permutations is even, as is the product of two odd permutations.

```
In[3]:= PermutationGroupQ[ Select[ Permutations[{1,2,3,4}],
  (SignaturePermutation[#]==1)&] ]
Out[3]= True
```

Are there more even or odd permutations, or the same amount for all n? Can you prove that each permutation has the same signature as its inverse? Why does the set of even permutations form a group, when the odd ones don't? How can the signature of a permutation be determined from its cycle structure?

3.5 Pólya's Theory of Counting

Pólya's theory of counting provides a tool for enumerating the number of distinct ways a structure can be built when symmetry is taken into account. The classic problem in Pólya theory is counting how many different ways necklaces can be made out of k beads, when there are m different types or colors of beads to choose from. The symmetry in the necklaces is reflected by the automorphism group of a simple cycle. Pólya's theory is developed in most combinatorics texts.

1. When two necklaces are considered the same if they can be obtained only by rotating the beads (as opposed to turning the necklace over), the symmetries are defined by k permutations, each of which is a cyclic shift of the identity permutation. When a variable is specified for the number of colors, a polynomial results.

```
In[1]:= Polya[Table[RotateRight[Range[8],i],
  {i,8}], m]
                      2     4     8
                 4 m + 2 m + m + m
Out[1]= -------------------
                  8
```

2. The number of colorings for unrestricted necklaces is determined by the automorphism group
of a cycle. There are approximately half as many
unrestricted necklaces as restricted ones, because
few necklaces are equivalent to their reverse.

```
In[2]:= Polya[Automorphisms[Cycle[8]], m]
                2     4     5     8
           4 m + 2 m + 5 m + 4 m + m
Out[2]=  -----------------------------
                       16
```

How large an m does it take for these functions to
be within close to a factor of two of each other? How
does it compare to the polynomial of a path instead of
a cycle? For $m = 4$, can you find the actual colorings?

3.6 Special Classes of Permutations

Involutions are permutations that are their own
multiplicative inverses. A permutation is an involution if and only if its cycle structure consists exclusively of fixed points and transpositions.

1. There are ten involutions on four elements.

```
In[1]:= Select[ Permutations[{1,2,3,4}], InvolutionQ ]
Out[1]= {{1, 2, 3, 4}, {1, 2, 4, 3}, {1, 3, 2, 4},
  {1, 4, 3, 2}, {2, 1, 3, 4}, {2, 1, 4, 3}, {3, 2, 1, 4},
  {3, 4, 1, 2}, {4, 2, 3, 1}, {4, 3, 2, 1}}
```

2. Since these are involutions, squaring them gives
the identity permutation.

```
In[2]:= Map[(Permute[#,#])&, %]
Out[2]= {{1, 2, 3, 4}, {1, 2, 3, 4}, {1, 2, 3, 4},
  {1, 2, 3, 4}, {1, 2, 3, 4}, {1, 2, 3, 4}, {1, 2, 3, 4},
  {1, 2, 3, 4}, {1, 2, 3, 4}, {1, 2, 3, 4}}
```

3. Involutions are exactly the permutations with a
maximum cycle length of two.

```
In[3]:= Map[ ToCycles, Select[ Permutations[{1,2,3,4}],
  InvolutionQ ] ]
Out[3]={{{1}, {2}, {3}, {4}}, {{1}, {2}, {4, 3}},
  {{1}, {3, 2}, {4}}, {{1}, {4, 2}, {3}},
  {{2, 1}, {3}, {4}}, {{2, 1}, {4, 3}},
  {{3, 1}, {2}, {4}}, {{3, 1}, {4, 2}},
  {{4, 1}, {2}, {3}}, {{4, 1}, {3, 2}}}
```

Derangements are permutations p with no element in its proper position, i.e., there exists no
$\{p_i = i, 1 \leq i \leq n\}$. Thus, derangements are permutations without a fixed point.

4. If a confused secretary randomly stuffs n different letters into n pre-addressed envelopes, what
is the probability that none of them end up where
they are supposed to? The ratio of the number
of derangements to the number of permutations
converges rapidly to $1/e$. Thus the answer is essentially independent of n.

```
In[4]:= Table[ N[ NumberOfDerangements[i]/(i!) ],
  {i,1, 10} ]
Out[4]= {0, 0.5, 0.333333, 0.375, 0.366667, 0.368056,
  0.367857, 0.367882, 0.367879, 0.367879}
```

5. In fact, rounding $n!/e$ gives a nicer way to compute the number of derangements.

```
In[5]:= Table[Round[n!/N[E]], {n,1,10}]
Out[5]= {0, 1, 2, 9, 44, 265, 1854, 14833,
  133496, 1334961}
```

Can you find a recurrence relation that counts the
number of derangements on n items? What kind
of procedure could be used to construct all derangements of n items?

A **binary heap** is a permutation p of length n such
that $p_i < p_{2i}$ and $p_i < p_{2i+1}$ for all $1 \leq i \leq n/2$. As a
consequence of this definition, p_1 must be the smallest
element of p. Heaps can be viewed as labeled binary
trees such that the label of the ith node is smaller
than the labels of any of its descendants.

6. Plotting the value of the ith element in a permutation against i provides a graphical way to
reveal its structure. In a random permutation,
there isn't much structure to reveal.

```
In[6]:= ListPlot[ RandomPermutation[127] ];
```

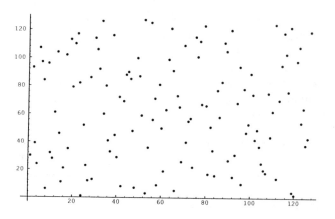

7. In a random heap, the first element is in the appropriate position, but the permutation gets progressively less ordered as we move to the right. Diagrams such as these illustrating a variety of different sorting algorithms appear in [15].

In[7]:= ListPlot[RandomHeap[127]];

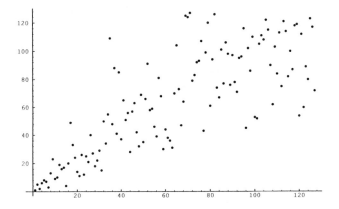

What do other permutations look like when plotted this way? How does the number of inversions effect the way the diagram looks? What about the number of cycles or runs?

4 Graph Theory and Related Problems

We have seen that a number of properties of permutations come to life when we can easily experiment with them. *Combinatorica* gives us that freedom, and can turn a curious student into a researcher—asking why a certain property always seems to hold. However, *Combinatorica* is even more powerful for exploring graph theory. Space limitations do not permit going into detailed examples, but we conclude with selected exercises from *Implementing Discrete Mathematics* [17] to whet your appetite. It does not take much imagination to construct additional interesting problems:

1. Experiment to determine the probability that a random tree has a center of one vertex versus two vertices.

2. Write a function to embed a hypercube in a grid graph, meaning to assign vertices of a hypercube to all the vertices of a grid graph such that the grid graph is contained within an induced subgraph of these hypercube vertices.

3. The **total graph** $T(G)$ of a graph G has a vertex for each edge and vertex of G, and an edge in $T(G)$ for every edge-edge and vertex-edge adjacency in G. Thus total graphs are a generalization of line graphs. Write a function to construct the total graph of a graph.

4. Use MakeGraph to construct the knight's tour graph for an $n \times n$ chessboard. This graph has a vertex for each square on the chessboard, with an edge connecting two squares that are separated by one knight move. Is this graph Hamiltonian for small values of n?

5. A **degree set** for a graph G is the set of integers that make up the degree sequence. Any set of positive integers is the degree set for some graph. Design and implement an algorithm for constructing a graph that realizes an arbitrary degree set.

6. Are there planar graphs that Brelaz's algorithm colors with at least five colors? If so, find an example of such a graph. Similarly, does Brelaz's algorithm always two-color bipartite graphs? Prove it does or find a counterexample.

7. Write a function to construct a random Euclidean graph as follows. Use RandomVertices to generate a list of n points in the unit square. Now construct a weighted complete graph where the cost of each edge is the distance between the embedding of the vertices.

Experiment with computing shortest-path spanning trees, minimum spanning trees, and traveling salesman tours on random Euclidean graphs. Are these subgraphs always planar embeddings?

8. Develop and implement a simple algorithm for finding the maximum matching in a tree, which is a special case of bipartite matching. Find a tree for which MaximalMatching does not find the maximum matching.

9. The **divorce digraph** is a binary relation associated with an instance of the stable marriage problem. The vertices correspond to the $n!$ matchings. There is a directed edge $\{a, b\}$ between matchings a and b if b results from an unstable pair in a leaving their respective spouses and

marrying each other, with the rejects getting paired off. Stable marriages are vertices with out-degree zero in the divorce digraph.

Write a function to construct the divorce digraph of a stable marriage instance. Verify that the following male $\{\{1, 2, 3\}, \{1, 2, 3\}, \{2, 1, 3\}\}$ and female $\{\{3, 2, 1\}, \{2, 3, 1\}, \{1, 2, 3\}\}$ preference functions result in a divorce digraph that contains a directed cycle. Further, show that whatever permutations represent male 1's and female 3's preference functions, the divorce digraph contains a cycle.

10. Perform experiments on the distribution of the length of the ith run in random permutations. How much longer, on average, is the first run than the second, and what value does the expected length converge to for large i?

11. If you look back on the n people of the opposite sex you have dated over the course of a lifetime, these people can be ranked from 1 to n in terms of desirability. Thus if these people are ordered by the time you started dating them, they define a permutation. We can use this model to decide how to marry the most wonderful person possible.

After you start dating person p_i, you must make a decision on whether to marry that person, or else reject him or her to start dating person p_{i+1}. We assume that anyone you ask will marry you unless you have previously rejected him or her, so there is no going back for seconds. The problem is how to decide whether the person you are dating is the best, or whether it is likely that someone better lies down the road. A good strategy is to date several people and reject them, just to get a feel for the distribution, and then accept the first person who is better than anyone you have seen. Experiment with strategies to maximize the chances of ending up with the best possible spouse, as well as strategies to maximize the expected value of your spouse.

12. Design and implement an efficient algorithm for constructing all partitions of n with distinct parts. Repeat, constructing all partitions of n into odd parts, even parts, distinct odd parts, and finally all **self-conjugate** partitions, meaning they are equal to their transpose.

13. Experiment with NumberOfTableaux to determine the shape which, for a given number of elements, maximizes the number of Young tableaux over all partitions of n.

References

1. Baxter, N., Dubinsky, E. and Levine, G. (1988). *Learning Discrete Mathematics with ISETL*, Springer-Verlag, New York.

2. Brown, M. (1988). *Algorithm Animation*, MIT Press, Cambridge, MA.

3. Brown, M. and Sedgewick, R. (1985). Techniques for algorithm animation, *IEEE Software*, 2:28–39.

4. Birgisson, B. and Shannon, G. (1989). Graphview: An extensible interactive platform for manipulating and displaying graphs, Technical Report 295, Computer Science Department, Indiana University, Bloomington, Ind.

5. Dean, N. (1991). An overview of NETPAD: Network Evaluation Toolkit for Processing, Analysis and Design, Technical report, Bellcore, Morristown, NJ.

6. Dao, M., Habib, M., Richard, J. P. and Tallot, P. (1987). "Cabri, an interactive system for graph manipulation," in *Graph Theoretic Concepts in Computer Science*, volume 246, New York, 58–67. Lecture Notes in Computer Science, Springer-Verlag.

7. Eades, P., Fogg, I. and Kelly, D. (1988). Spremb: A system for developing graph algorithms, Technical report, Department of Computer Science, University of Queensland, St. Lucia, Queensland, Australia.

8. Fajtlowicz, S. (1987). "On conjectures of Graffiti II," *Congressus Numerantium*, 60, 187–197.

9. Fajtlowicz, S. (1988). "On conjectures of Graffiti," *Discrete Mathematics*, 72, 113–118.

10. Gonnet, G. (1984). *Handbook of Algorithms and Data Structures*, Addison-Wesley, London.

11. Himsolt, M. (1990). $graph^{Ed}$: an interactive tool for developing graph grammars, Technical report, Universität Passau, Passau, West Germany.

12. Krishnamoorthy, M., Spencer, T., Echeandia, M., Faulstich, A., Kyriazis, G., McCaughrin, E., Maroulis, C. and Pape, D. (1990). Graphpack: a software system for computations on graphs and sets, Technical Report 90-7, Rensselaer Polytechnic Institute, Troy, NY.

13. McKay, B. (1990). Nauty user's guide, Technical Report TR-CS-90-02, Department of Computer Science, Australian National University.

14. Nijenhuis, A. and Wilf, H. (1978). *Combinatorial Algorithms* (second edition), Academic Press, New York.

15. Sedgewick, R. (1988). *Algorithms* (second edition), Addison-Wesley, Reading, Mass.

16. Siegel, M. (1988). "The use of computers in teaching discrete mathematics," in D. Smith, G. Porter, L. Leinbach, and R. Wenger, (eds.), *Computers and Mathematics: The Use of Computers in Undergraduate Instruction*, MAA Notes number 9, Mathematical Association of America, Washington, D.C., 43–46.

17. Skiena, S. (1990). *Implementing Discrete Mathematics: Combinatorics and Graph Theory with Mathematica*, Addison-Wesley, Redwood City CA.

18. Stasko, J. (1989). Tango: A framework and system for algorithm animation, Technical Report CS-89-30, Department of Computer Science, Brown University, Providence, RI.

19. Stanton, D. and White, D. (1986). *Constructive Combinatorics*, Springer-Verlag, New York.

20. Wolfram, S. (1988). *Mathematica*, Addison-Wesley, Redwood City, Calif.

Use of Symbolic Computation in Probability and Statistics

Zaven A. Karian
Denison University

Andrew Sterrett
Denison University

1   Introduction

During the six years or so that symbolic computation has been used in undergraduate mathematics instruction, most of the effort has been directed to improving calculus instruction.

We have sensed all along, without explicit acknowledgement, that if our initiatives for improving calculus instruction through symbolic computation succeeded, we would have to consider the use of the technology in other parts of our curriculum. Once students have been introduced to a computer algebra system (CAS), the use of a CAS in advanced courses will be unavoidable. Students will insist on using symbolic computation, and rightly so, in their advanced courses. It makes little sense to use a CAS to relieve the drudgery of techniques of integration in calculus and refrain from using it in Differential Equations. It is reasonable to expect that in the next few years there will be many new initiatives directed at using symbolic computation in upper-level mathematics courses. The issue is not **if** we will use CASs in advanced courses, it is **how** we will use it to promote better learning by our students.

The most obvious upper-level courses where CAS can be used effectively are linear algebra, differential equations, and numerical analysis. We are aware of a number of initiatives in linear algebra and in differential equations, some of which are described in Part III of this volume. It may be less obvious, but we believe potentially as valuable, to also use CASs in probability and statistics, abstract algebra, and discrete mathematics.

At Denison University, we began to use a CAS (*Maple* in our case) in some advanced courses two years ago. What we will describe here is some of what we have done in our two-semester mathemat-ical probability and statistics sequence. The course that we teach is a reasonably standard one, taught from Hogg and Tanis [1]. Although computation and interpretation of various statistics and development of the ability to deal with large amounts of data are among its goals, the primary aim of our two-semester sequence is to develop an understanding of the basic concepts of probability and statistics.

One way of making abstract ideas more comfortable and more concrete is to have students look for patterns by investigating various alternatives. This idea was being used before the general availability of CASs (see, for example, Sterrett and Karian [4] and Tanis [5] and [6]). However, with the power of modern CASs we can engage students more effectively in the exploration of the relationships between sample statistics and the parameters of distributions.

The more powerful CASs (at least this is the case for *Maple*) have built-in features for generating random samples from well-known distributions. Unfortunately, the statistical package embedded in *Maple* lacks certain essential features (e.g., histogram plotting so that students can quickly get a sense of the shape of a distribution or a sample). We have remedied these problems by adding 25 routines to *Maple*, mostly plotting and special generation routines. With these additions, we now direct our students' attention to some of the basic concepts of the course. A list of the routines that we have added is given in the Appendix.

2   Approximations to Discrete Distributions

Many of the discrete distributions (e.g., binomial, Poisson, hypergeometric) studied in typical probability and statistics courses can be approximated by a normal distribution, $N(\mu, \sigma^2)$, with suitably chosen mean, μ, and variance, σ^2. Through special assignments, students are be directed to discover the approximation properties of $N(\mu, \sigma^2)$ prior to a formal classroom discussion of these issues. Here is how this can be done in the case of $B(n, p)$ (the binomial distribution associated with n independent success-failure trials with probability of success p).

Exercise 1.

1. Study the shapes of the binomial distribution by obtaining histograms for $B(n, p)$ for various values of n and p. Does it look like a normal distribution could approximate certain binomial distributions? Which ones?

2. If you wanted to approximate $B(n, p)$ by $N(\mu, \sigma^2)$, what μ and σ^2 would you choose? Superimpose the graphs of the $N(\mu, \sigma^2)$ p.d.f.'s that you chose on their corresponding $B(n, p)$ histograms. How do the approximations look?

3. Suppose we want to find $P(2 < X \le 6)$ where X is a $B(15, 1/3)$ random variable. By looking at a graph similar to those considered in Part 2, you can see that this probability can be approximated by $\int_a^b f(x)\, dx$ where $f(x)$ is an appropriately chosen normal p.d.f. To obtain this approximation, what values of a and b should be used? Find $P(2 < X \le 6)$ and compare it to your approximation.

4. Obtain histograms for $B(n, p)$ for $p = 0.1$ and $n = 5, 10, 20, 35$ and 50. Does it look like a normal distribution could approximate certain $B(n, 0.1)$ distributions? Make some general observations relating n and p to the quality of approximations.

It should be noted that it takes relatively little effort to obtain the superimposed graphs mentioned in Part 2. For example, the *Maple* sequence[1]:

```
binom:= BinomHist(15,1/3):
norm := NormalPDF(4,8/3):
plot({binom, norm}, -4..15, style=LINE);
```
will produce the graph shown in Figure 1.

Since the binomial and normal distributions have already been studied by the time this exercise is assigned, students know that the mean and variance of

---
[1]**BinomHist** and **NormPDF** are from our supplement and not part of *Maple*.

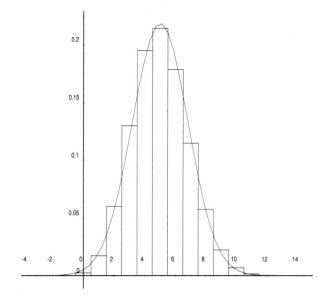

Figure 1: A normal approximation to $B(15, 1/3)$.

$B(n, p)$ are np and $np(1 - p)$, respectively. Hence, they generally have no difficulty in choosing μ and σ^2 correctly in Part 2. Part 3 gives students the opportunity to discover the need to use a "continuity correction" in their approximation. This is reinforced when approximations to other discrete distributions are taken up in similar exercises later in the course.

3 Sampling Distributions

Consider, for example, the idea of obtaining empirical evidence for

$$\mu_{\bar{x}} = \mu \quad \text{and} \quad \sigma_{\bar{x}}^2 = \sigma^2/n$$

for the sampling distribution of means. Conceptually, it is easy to decide how to go about this. Computationally, it could be quite complicated: an underlying distribution must be specified, random samples of various sizes must be extracted from this distribution, and the sample statistic $\bar{x}$ must be computed for each sample.

Exercise 2. Study the sampling distribution of the means of samples from the normal distribution with mean $\mu = 100$ and variance $\sigma^2 = 144$, $N(100, 144)$.

1. If $\bar{x}_1, \bar{x}_2, \ldots, \bar{x}_{100}$ are 100 means of samples of size n, is there a discernible distribution that fits $\bar{x}_1, \bar{x}_2, \ldots, \bar{x}_{100}$? Answer this question by generating sets of $\bar{x}_1, \bar{x}_2, \ldots, \bar{x}_{100}$ and looking at such

things as the averages, variances, histograms, and empirical p.d.f.s. of $\bar{x}_1, \bar{x}_2, \ldots, \bar{x}_{100}$.

2. In a similar way study the sampling distribution of the medians of samples from $N(100, 144)$.

3. Do the same for sample variances.

4. Does $\bar{x}$ seem to be an unbiased estimator of μ? Does the sample median seem to be an unbiased estimator of μ? Does the sample variance, s^2, seem to be an unbiased estimator of σ^2?

5. Of the two estimators (sample mean and sample median) of μ, is one better than the other? If so, in what sense is it better?

6. Would your answers to Part 4 change if we had started with a distribution that was not normal? Investigate some other cases and justify your answer.

This exercise actually consists of two exercises that have been combined here because of the flow of ideas from sampling distributions to unbiased estimators. Parts 1, 2, and 3 are assigned when the sampling distribution of $\bar{x}$ is first introduced. Notice that although Part 3 can be empirically investigated at this stage, the results and the pattern sought may not be that obvious to the students. It is nevertheless important for them to see empirically that an "unusual" shape can occur.

Parts 4, 5, and 6 are assigned as the idea of unbiased estimators is introduced and before any discussion of minimum variance estimators. Most of our students are able to discover the connection between the symmetry of the underlying distribution and the unbiasedness of the sample median as an estimator of μ. Fewer students, about half, observe that, generally, $\sigma_{\bar{x}}^2 < \sigma_{\text{median}}^2$, and consequently $\bar{x}$ is a better estimator as a result of smaller variability. These ideas can lead, in a natural way, to the use of bootstrapping methods for parameter estimation (for a discussion see Moore and Witmer [3]) in situations where the theory is not well-developed.

The *Maple* statistical package, together with our supplement, allows students to consider many sets of $\bar{x}_1, \bar{x}_2, \ldots, \bar{x}_n$ in order to look for patterns. For example, the following *Maple* interaction[2] will give information about 100 means of samples of size 20 and plot an empirical p.d.f. for $\bar{x}_1, \bar{x}_2, \ldots, \bar{x}_{100}$.

---

[2]**NormMeanS** and **EPDF** are from our supplement and not part of *Maple*.

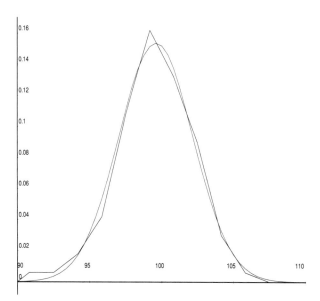

Figure 2: Empirical p.d.f. with approximation.

```
X   := NormMeanS(100,144,20,100):
ave := average(X);
            99.60297730
var := variance(X);
            6.742421962
pdf := EPDF(90,110,12,X):
plot(pdf, style=LINE);
```

If the student observes, from the values of **ave** and **var**, that the distribution of $\bar{x}$'s is approximately $N(99.6, 6.7)$, then $\mu_{\bar{x}} = \mu$ becomes an easy observation for students, but $\sigma_{\bar{x}}^2 = \sigma^2/n$ is much more subtle.

```
y := NormalPDF(99.6,6.7)
```
followed by
```
plot({y,pdf}, 90..110, style=LINE):
```
can be substituted for the previous **plot** command to give the graph shown in Figure 2. At this point, students are asked to vary the sample size and make another attempt to discover the result for variances.

4   Statistical Testing

Most junior-senior level books on probability and statistics discuss the role of $\alpha = P(\text{type I error})$ in constructing statistical tests. Since $\beta = P(\text{type II error})$ generally cannot be given in explicit form, it receives considerably less attention. The power function of the test (or its graph) provides a great deal of information about β for various values of the parameter being tested. However, since power functions are difficult to plot, most books give only one or two examples/exercises related to the power function. Through the use of *Maple*, we ask our students to plot and discuss the power function in all statistical testing problems.

Exercise 3. Let $X_1, X_2, \ldots, X_{25}$ be a random sample from $N(\mu, 4)$. Consider the following 3 tests for testing $H_0 : \mu = 0$ against $H_1 : \mu \neq 0$.

Test 1: Reject H_0 if $\bar{x} > 0.658$
Test 2: Reject H_0 if $\bar{x} < -0.658$
Test 3: Reject H_0 if $|\bar{x}| > 0.784$.

In each case determine α, the probability of Type I error, and study the power function of the test. What do you conclude about the suitability of these tests for various values of μ?

In all three tests $\alpha = 0.05$. However, the quality of the test for various values of μ changes dramatically from one test to another. Most undergraduate texts use the power function only in the simplest situations because of the difficulties associated with plotting these functions. These graphs, easily obtainable through *Maple*, shed considerable light on the qualitative differences of the three tests. For those interested, after loading the statistical package by issuing the `with(stats):` command, the following *Maple* interaction produces a graph of the power function for Test 1. A plot of the power functions for all three tests is given in Figure 3

```
a := 5 * ( 0.658 - mu ) / 2;
powerf := 1 - N(a);
plot ( powerf, mu = -1.5 .. 1.5 );
```

We also use a modification of Exercise 3 to have students investigate the power functions of tests based on a fixed sample size, but with varying levels of significance. A third variation on this theme is to keep a fixed significance level and study the effect of sample size on the power function of a test.

5   Confidence Intervals

In the study of confidence intervals for parameters of various distributions, we typically ask students to find intervals that have a given probability, $(1 - \alpha)$, of containing a parameter. Formulas are derived, based in part on statistics obtained from a random sample, and then used to determine the endpoints of confidence intervals in a variety of situations (e.g., for p in $B(n, p)$, the mean of a normal distribution, etc.).

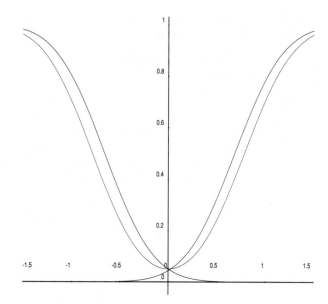

Figure 3: Power functions for the tests of Exercise 3.

Students get a deeper appreciation of confidence intervals when they see some evidence that the intervals, in fact, contain the parameter being estimated $100(1 - \alpha)\%$ of the time. The idea of providing such evidence is simple: obtain many (20 or more) random samples; in each case produce a confidence interval; and then determine what proportion of the intervals contains the parameter. In practice, doing this without the aid of a statistical package or a symbolic computing system would be impractical, if not impossible.

Exercise 4: Use *Maple* to generate 25 samples, each of size 10, from $N(100, 225)$. In each case, produce an 80% confidence interval for the mean, μ. How often does the confidence interval contain μ?

In this exercise an 80% confidence interval is used, instead of the more customary 90% or 95% one. This makes it more likely that each student will encounter intervals that do not contain μ. Of course, few of them will get exactly 5 (the expected number) intervals that do not contain μ. After students have done this exercise, they each have evidence (based on 25 samples) of how often their intervals contain μ. An in-class discussion can take advantage of the individual student results by considering the accumulated result of the entire class. Even in a small class of size 10, the accumulated result would be based on 250 samples, and approximately 50 of the 250 intervals will not contain μ.

The following *Maple* interaction[3] shows the computation of a single confidence interval.

```
c := 1.282*15/sqrt(10):
X1 := NormalS(100,225,10):
ave1 := average(X1):
LeftEP := ave1 - c;
             99.46571509
RightEP := ave1 + c;
             111.6088613
```

The second, third, fourth, and fifth command lines will have to be repeated to obtain additional confidence intervals. This process can be shortened by defining functions that compute `LeftEP` and `RightEP`, the left and right endpoints, respectively. The more enterprising students figure out how to put these statements in a *Maple* loop and avoid giving repetitive commands.

As in most textbook problems, we assumed in this exercise that the underlying distribution was known; moreover, that it was $N(100, 225)$. However, in most applications, the investigator has no information about the distribution from which the sample was taken. How would one find an interval estimate for the population mean (or some other parameter) if all one has is a random sample? In such realistic applications a CAS could be used to implement a bootstrapping method that involves repeated samplings (see Karian and Dudewicz [2], pp. 279–282) to find a confidence interval.

References

1. Hogg, R. V. and Tanis, E. A. (1988). *Probability and Statistical Inference*, Macmillan Publishing Company, New York.

2. Karian, Z. A. and Dudewicz, E. J. (1991). *Modern Statistical, Systems, and GPSS Simulation*, W. H. Freeman and Co., New York.

3. Moore, T. L. and Witmer, J. A. (1991). "Statistics Within Departments of Mathematics at Liberal Arts Colleges," *The American Mathematical Monthly*, V. 98, No. 5.

4. Sterrett, A. and Karian, Z. A. (1978). "A Laboratory for an Elementary Statistics Course," *The American Mathematical Monthly*, V. 85, No. 2.

5. Tanis, E. A. (1974). "Mathematical Probability and Statistics Computer Laboratory," *Int. J. Math. Educ. Sci. Technol.*, V. 5.

Appendix

This Appendix lists the routines that were added to *Maple*, together with an indication of how they are called during a *Maple* session.

p.d.f. Generators

1. NormalPDF(μ, σ^2) — normal distribution.
   ```
   y:=NormalPDF(10,16);
   ```
   will make $y(x)$ the $N(10, 16)$ p.d.f.

2. CSPDF(r) — chi square distribution.

3. ExponPDF(θ) — exponential distribution.

4. GammaPDF(α, θ) — gamma distribution.

Sample Generators

1. UniformS(a, b, n) — random sample of size n from $U(a, b)$.
   ```
   A:=UniformS(10,40,100);
   ```
   will make A a random sample (actually a list) of size 100 from $U(10, 40)$.

2. CauchyS(n) — Cauchy distribution.

3. LogisticS(n) — logistic distribution.

4. ExponentialS(θ, n) — exponential distribution.

5. NormalS(μ, σ^2, n) — normal distribution.

6. BVNS$(\mu_X, \sigma_X^2, \mu_Y, \sigma_Y^2, \rho, n)$ — bivariate normal distribution.

Samples from Sampling Distributions

1. UnifMeanS(a, b, n, m) — m means of random samples of size n from $U(a, b)$.
   ```
   A:=UnifMeanS(10,40,20,100);
   ```
   will make A a sample (actually a list) of 100 means of random samples of size 20 taken from $U(10, 40)$.

2. UnifMedianS(a, b, n, m) — medians from uniform distribution.

3. UnifSumS(a, b, n, m) — sums from uniform distribution.

---

[3]`NormalS` is from our supplement and not part of *Maple*.

4. NormMeanS(μ, σ^2, n, m) — means from normal distribution.

5. NormMedianS(μ, σ^2, n, m) — medians from normal distribution.

6. NormVarS(μ, σ^2, n, m) — variances from normal distribution.

7. NormSumS(μ, σ^2, n, m) — sums from normal distribution.

8. NormTransVarS(μ, σ^2, n, m) — $(n - 1)S^2/\sigma^2$ from normal distribution.

9. ExponSumS(θ, n, m) — sums from exponential distribution.

Histogram Plots

1. BinomHist(n, p) — histogram for binomial distribution.

            a:=BinomHist(n, p):

   followed by

            plot(a, style=LINE);

   will produce a histogram of the $B(n, p)$ distribution.

2. PoissonHist(μ) — histogram for Poisson distribution.

3. Hist(A) — histogram for the list A. If A is a list,

            h:=Hist(A):

   followed by

            plot(h,style=LINE);

   will produce a histogram of the data in A.

4. HIST(min, max, n, A) — histogram for the list A. If A is a list, $min \leq$ minimum entry of A, $max \geq$ maximum entry of A, then

            h:=Hist(min, max, n, A):

   followed by

            plot(h,style=LINE);

   will produce a histogram with n intervals of the data in A.

Empirical p.d.f. Plots

1. Epdf(A) — empirical p.d.f. based on the list A. If A is a list,

            a:=Epdf(A);

   followed by

            plot(a, style=LINE);

   will produce a plot (an approximation) for the distribution of A.

2. EPDF(min, max, n, A) — p.d.f. for the list A. If A is a list, $min \leq$ minimum entry of A, $max \geq$ maximum entry of A, then

            a:=EPDF(min, max, n, A):

   followed by

            plot(a, style=LINE);

   will produce a plot (an approximation) for the distribution of A.

Part V:

Getting Started and
Review of the Literature

Reformation at St. Olaf

Paul D. Humke
St. Olaf College

1 A Bit of Background

Interest in calculus reform at St. Olaf is quite ancient, predating my arrival by at least eight years. Indeed it was the success of the original St. Olaf "*year and out calculus*"[1] course which attracted me to visit the College for the 1979–80 academic year. To be honest though, the effervescence and persistence of Ted Vessey were the main reason I came. After having been at large state universities all my adult life, the inner workings of the St. Olaf Mathematics Department were nothing short of miraculous to me. The energy of the place is extraordinary and the atmosphere electric with the activity of our profession. This is important to understand for the discussion to follow. "Calculus reform" has been only one of several major themes at St. Olaf during the past decade and although it has been an important project here, it has not been a dominant theme. It is one of the things we deal with, but it has been for some time now.

2 Why This Current Round?

Arnie Ostebee and I came the same year (I was visiting, Arnie was a new assistant professor) and Paul Zorn a year later. New folks are encouraged to get involved immediately and it was perhaps natural we began by discussing the St. Olaf curriculum. The waters had just begun to calm after changing calculus to a one-year course and the next step was to review the contents of linear algebra; calculus appeared secure for decades. Linear algebra was and is the pivotal course in our curriculum for it is in this course

that our majors are forged. But partially because calculus was now only a one-year course, the nature of linear algebra needed to change. Particularly, it was important to put a heavier emphasis on geometric aspects (low dimensional and Euclidean) as our one-year calculus had become almost entirely a single variable course and devoid of analytic geometry. About this time, Tom Banchoff came to visit and spoke to a hall packed to the rafters with students (and faculty) wanting to learn of the geometry of the 4th dimension. Over 350 souls attended and many were saved that day.[2] Among those was a group of my linear algebra students who wanted to reproduce Banchoff's 4-dimensional cube film and this we did on primitive equipment the next semester. I asked Arnie how much it would cost to give all our linear algebra students similar opportunities. It seemed natural to want to take advantage of this excitement about mathematics. Of course, it wasn't possible then; the cost was prohibitive. Arnie suggested that calculus was a better place to start. Paul Zorn had some experience with SMP[3] and the three of us, together with several others, began to investigate the idea of allowing students to mull over the ideas of calculus in an environment that did elementary calculations for them.

3 What We Wanted to Do

What I wanted to do was to deliver the excitement of calculus to more of my students. This was all, but the notion entails more. As a student, I found the excitement of calculus to be in the ideas; calculus was powerful. When deciding on my own undergraduate major, I determined that I thought about other

---

[1] The idea of this course is to strip away essentially all the multivariable material and teach a year of one-dimensional calculus. The multivariable material then follows linear algebra, but is not required of all majors.

[2] We talk like this at St. Olaf.

[3] A precursor of *Mathematica*

disciplines via mathematics and my choice of majors was corollary to this observation. Calculus was the liberator. It was clear to Paul, Arnie, myself, and others, that calculus had become very topic oriented as contrasted to the idea-oriented course we'd had as students. Worse, calculus exercises had become so routine and algorithmic that even students who were very successful on assignments and tests could know nothing of the stuff of calculus.

It was also clear that we could no longer teach the calculus courses we had taken as students. Several of us had used Spivak! Most of us had had impressive amounts of solid geometry and sophisticated precalculus algebra as precursors to our calculus courses. We couldn't treat calculus in our classrooms as our teachers had because the students had changed in substantial and important ways. Someone suggested we "bite the bullet" and teach a sophisticated calculus course in order to send a message to the high schools. I'm afraid the messages sent would be to us, first from our students and then from our administration. In any case, most of us are more interested in mathematics than sending messages to anyone. Paul pressed the point that machines could handle routine computations and perhaps we could use this facility to teach a deeper calculus course. I was interested primarily in the ability of the machines to graph and more importantly, to allow students to investigate graphs. Arnie was interested in all of this; too, he knew that machines were now available at a price within hope of paying. At this point, we formalized goals for our calculus course, formed our strategy for implementing these goals, set timetables, wrote grants, and negotiated with granting agencies. In short, the real work was about to begin. It is important to mention that all ideas, preliminary and otherwise, were discussed, altered and evaluated in open department meetings. No grants were applied for, no programs begun, no curricula changed without departmental consensus.[4]

4   What We Did

It would have been too risky to change our entire curriculum as radically as we proposed without any experience either by St. Olaf or anywhere else for that matter. We decided to begin with one experimental class taught by Paul Zorn in the January term of 1986–87 followed by two small experimental sections of second term calculus in the spring. Paul taught one of these,

Cliff Corzatt the other. A first round of course materials was written prior to the January course and rewritten several times during each of the next three years. During the fall of 1987–88, we were to teach three sections of regular calculus I and one honors calculus I section using the new materials. But, prior to this, Paul and I left! I, for Budapest to do research in real analysis and Paul to Purdue to pursue several complex variables. We didn't anticipate any major problems since all faculty had seen the materials and most had attended a locally-run workshop or two on the use of *SMP*.

The faculty teaching special fall sections had attended (and been paid to attend) a week-long training session and all had written some of their own materials. Fundamental to our project was the cascade of faculty involvement in teaching, editing, and decision making. Faculty would be rotated through the program during the next three years so that at the end of four years, 15 of our 24 faculty would have taught using the new materials. Grants, some seven in all, paid for release time, equipment, training, a conference or two, and much more. So despite sabbatical leaves for the two Pauls, confidence ran high.

5   What Happened

I returned from Budapest and Paul came from Purdue to attend the January meeting in Atlanta. Paul, Arnie, and I had a long breakfast to discuss how the first full year of the program was going. I'll never forget that breakfast! An unbelievable effort had been put into getting everything ready for the students. Arnie had battled legions of hardware and software sellers who promised but didn't want to deliver. Paul's materials had been refined and re-refined and countless hours had been spent getting everything to work properly. Physically and intellectually we were as prepared as humanly possible. Yet, I believe "hate" is not too strong a word to describe how students received our efforts. And for faculty used to being loved by students, "hate" is just awful. Our first objective evaluations came from our students and we had much to learn. As with any evaluation, what is written is often raw and subjective and can hurt. After reflection, here are some things we learned early in the game.

5.1   Curricular Lessons

We tried to do too much. We attempted to cover all the old syllabus topics in a new way. Namely, we asked the students to think. Thinking takes copious amounts of time and we hadn't allowed for this. We'd hoped to teach an ideas course that covered all the topics of a traditional course; it isn't possible.

We attracted a new breed. Our new course seemed[5] to turn off many students who would have been attracted to mathematics in a more traditionally taught course. Those students who had learned to succeed in mathematics by using mimicry had trouble adjusting to our approach. Many of these students were very bright but had developed successful yet ultimately dysfunctional methods with which to study mathematics. But we also seemed to attract students who wouldn't have found calculus attractive previously. The gains versus losses were not encouraging at first; it seemed we were losing more than we were gaining. Alarmingly, it seemed that we were turning off groups we wanted to attract to mathematics and that we had been successfully attracting in the past, specifically, women. Subsequent studies of student enrollment patterns revealed a more complex situation some aspects of which are described below.

Students in the new sections were more likely to continue studying mathematics. Although a significantly higher percentage of students from the new sections dropped out of calculus initially, it also appeared that those who stayed in the course were significantly more likely to take advanced courses in mathematics. These students were also more likely to declare a major in mathematics.[6] Because of our concern about discouraging our women students we looked at gender filtered numbers as well. The results seemed consistent; that is, a larger percentage of women dropped out initially, but those who remained were more likely to take advanced courses in mathematics and were also more likely to declare a mathematics major.[7]

Teachers were disillusioned. This was perhaps the most difficult blow to absorb. We have effective, professional, dedicated, revered teachers in our department. Most never had had negative ratings

from students [8]. To experience student hostility (and for the first time) feels bad, very bad. Some serious soul searching went on, perhaps especially with Paul, Arnie, and me.

Students avoided the marked sections. The grapevine at St. Olaf has been fiberoptic for as long as I've been here. Students heard about the "special sections" and wanted no part of them. Several went so far as to announce they would forgo taking calculus rather than sign up for one of the new sections.

5.2 Pedagogical Lessons

Students felt "this is not mathematics." I must admit that this criticism was totally unexpected for me. Nonetheless, it was a widely held student opinion. For a few, our approach was a marvel; they were not interested in "the old mathematics, but this..." But most of our students come to St. Olaf having been successful at mathematics throughout their school careers. They had developed successful strategies for "learning mathematics" and were confident in their abilities. We handed them confusion and they resented it. They knew that every "reasonable" problem could be solved in less than a minute if one knew the template, and less than two if one had to refer to the template conveniently located in the section prior to the exercise. We, on the other hand, gave problems that were intricate and ill defined. For our students, a mathematics problem always had a **right** answer. Indeed, many of our brightest students were/are attracted to mathematics for precisely this reason! We, however, gave problems where there was neither a correct answer nor even a correct methodology. For our students, this was unexpected; this was not mathematics! Corollary was the fact that the teaching was bad—"I know what math is, it's always been that, my math teachers were always able to make things clear, even easy up to now, therefore ...".

Student misconceptions were revealed. Perhaps this is the case whenever one presents material in a different manner, but we were stunned by the low level of understanding our students had of even very fundamental concepts. I'll give an extended example and let the reader's imagination supply others.

The students, our wonderfully prepared St. Olaf students, had only the vaguest notion of what function meant, and this limited vision surfaced in many

---

[5]The word *seemed* is to mean these are personal impressions. We collected a good deal of data about what the dynamics were in our calculus courses, and it's my opinion about what this data says that I'm recounting in this paragraph.

[6]Information about how many of these students actually completed a major is not yet available.

[7]The Office of Educational Research at St. Olaf was very helpful in helping us evaluate our program.

---

[8]See introduction.

contexts. Most often, "function" to our students meant something akin to an expression of the form, "y = polynomial in x." Most were able to supply an appropriate equation to the demand "give me a trigonometric function" or the question "what is an example of a logarithmic function?", and even sketch a correct graph of these, but this was the extent of their knowledge. A log table was a table not a function. Force was not a function of mass and acceleration in any meaningful sense other than the rote "$F = ma$." Students did not understand phenomena in terms of mathematics; rather, mathematics was simply an independent set of facts for them.

Too, this "limited vision" showed up when students tried to learn *SMP* syntax; the importance of parentheses or brackets was not clear to them and the root of this insensitivity seemed to point to a failure to broadly understand what is meant by function. Even the notion of linear function was understood only in the most superficial way. Students would agree that $y = 3x + 4$ was a linear equation and y was a function of x. However, most would agree that x was not a function of y. When given a table of values for x and y reflecting the relationship $y = 3x + 4$, most students would recognize the relationship as linear because they would first derive the equation $y = 3x + 4$ and then see it's a linear equation. Few recognized a table reflecting $y = \pi x + e$ as linear or even saw that an exponential table is not linear. They couldn't find the equation! To be honest, our honors sections had a much fuller understanding of the notion of function than our non-honors sections, but for some regular calculus students (used to getting A's and B's) such shallow roots were debilitating.

Lecturing can reinforce bad habits. In topic- or technique-oriented classes, lecturing can be very effective. The students need facts, the teacher provides them and shows why they are facts. Certainly efficient. But in a conceptual course[9] students are asked to investigate situations using what mathematical tools they have. They must not be left entirely to their own devices or little learning will result. Mistaken thinking needs to be shown for what it is, hints need to be given, overlooked information needs to be pointed out and questions need to be formulated and

reformulated: all of this in order that the investigation be fruitful and lead to significant mathematics. It became apparent that if we wished to have students engage calculus at a deeper level, more was demanded of us. Lecturing, even superb lecturing, was neither necessary nor sufficient and often reinforced an unholy alliance between teacher and student; the teachers job was to show how to do "it," the students job was to mimic "it" until "it" could be done without reference to notes, and then a test would be given to check "it." Such an alliance is perhaps appropriate if one is interested in teaching and learning the facts of mathematics, but not if learning to think mathematically is a primary goal.

Some things went as planned. It's easy to dwell on the unexpected; after all, these had the greatest effect on us. Some things went as we planned and deserve some print. I do believe students learned that integration and antidifferentiation are different and the fact that under some circumstances they're equivalent receives its proper due. Estimation becomes a legitimate enterprise and students learn to use quick and dirty estimates to check results. They also learn that although exact answers are preferred, approximate answers with error estimates are acceptable and sometimes the best one can do. Students understand what it means for an infinite series to converge and have some notion of how a series can diverge. Rates of convergence or divergence of positive term series play a prominent and meaningful role for students. Convergence of sequences and series of functions appears to be easier conceptually than convergence of series of numbers at the level we teach each topic.[10]

6 Our Response

As with any cornered animal, we responded to these lessons frequently and with great vigor.

6.1 We revised the syllabus

The revamping of our calculus syllabus was certainly the most fascinating of these responses. It was generally held that the amount covered in calculus needed to be reduced by about 30% and in a department as congenial as ours, this wasn't a difficult task, it was impossible. Literally. In department meeting after department meeting, candidate topics were brought

---

[9]It seems difficult to make clear the concept of conceptual. It does not mean "proving every theorem." It does not mean a "touchy-feely" approach to mathematics. To me it means, for example, we're interested in the concept of, say length. First, how does one find it. Which objects have length, which don't. How could one define it? etc.

[10]To some this was a surprise.

forth and each was found to have qualities so redeeming that faculty were willing to take the ax in their stead. In the end, we were not even able to cut partial fractions![11] We failed miserably. But after a year of intermittent discussion, we decided to write a new syllabus from scratch. The zero-based syllabus creation succeeded. In a short time a syllabus was compiled for each of our two semesters. These syllabi are quite lean but they make mathematical sense and leave sufficient time for exams, review days, and instructor discretion. They are not chock-a-block with the nuggets of calculus; they describe a main channel leaving time to stop to explore an eddy or two.

6.2 We revamped the student notes

We spent a good deal of time and energy revamping the class notes we gave to students. Al Magnuson wrote a day-by-day set of first-semester notes and made them available to everyone. I used many of Al's ideas and much of his structure in writing a detailed set for myself. Many others did the same. All of our materials were available electronically to one another and this was very helpful. After our first experiences, we were better prepared for our students and had some notion of how they would react to our materials. All of us reintroduced routine exercises. My own method is to have two or three assignments of a relatively routine nature followed by a single, open-ended, often ill-posed problem which the students have three or four days to work on. The routine exercises are good for confidence building and class morale.

6.3 We took extra care with the hardware and software

Nothing seemed as deadening to a class as a glitch in the hardware or software. Already skeptical of their own abilities to learn mathematics using computer tools, students readily use technical problems as a lightening rod for their own fears and frustrations. After our year-long shakedown though, the number and intensity of technical problems decreased rapidly.

6.4 Faculty training was broadened and intensified

Our project emphasized a cascade of faculty using a computer algebra system in calculus, coupled with appropriate faculty training. The nature of faculty training changed in both content and focus because of experiences during the first year. The first training session, held during the second week of August 1986, emphasized familiarity with hardware and software and practice in creating open-ended problems. In addition to these topics, subsequent training included several discussions on student reactions, the importance of routine exercises, student expectations, and, of course, testimony as to what worked and what did not. It is difficult to overestimate the importance of these week-long sessions to the success of our project. I won't continue the attempt.

What was the effect of our response? It's absolutely impossible to say, but things are going much more smoothly now. I've taught six calculus courses using materials either I or others wrote as part of our project. The use of a computer algebra system is inherent in almost every (about 2/3) assignment I give. Student reaction to my classes has been good, even enthusiastic, and student opinion of me seems quite resilient to change. I feel the calculus students in my project classes are different in several important ways.

1. They think like mathematicians.[12]

   By this I simply mean that when confronted with a problem they try to think. They ask themselves questions: "What do I know about this?" "What is being asked?" "What might that mean?" And about the problem they ask, "What's going on here, anyway?"

2. They're willing to try anything.

   No longer do students simply look at Section 5 techniques to solve Section 5 problems. They'll use **anything** to get a handle on a tough problem and have often surprised me with elegant and elementary solutions to problems I thought required a bit more machinery.

3. They're far more insecure about what they've learned.

   As I see it, this is corollary to point 1 above. Such insecurity is inherent in doing real mathematics. The students are used to feeling confident about their mathematical abilities, used to knowing which problems they should be able to solve, used to dominating the material. In fact, the security they've felt in the past was not well founded and the gap between what they expect to

---

[11]I myself went to bat for partial fractions!!

[12]Wow, is this a difference!

learn from me and what I expect them to learn is not inconsequential. I've spent many hours dealing with this expectation gap. It's a problem.

My practical answer to the request to compare pre-project and project calculus teaching is "The good Lord willing, I'll never have to teach the old way again." When asked what the main difference in my course is I respond: "I feel the students are engaging the real stuff of mathematics now." When asked about which computer algebra system I recommend my answer is: "at present such computer algebra systems as *Maple* or *Mathematica* are at the bare minimum of what is needed." [13]

Although there is a good deal of disagreement in the department about how calculus *should* be taught, there is general agreement on our calculus syllabus. This syllabus is very lean and focuses on the core ideas of the single variable theory; it is, in many ways, a remarkable document. Calculus at St. Olaf is a far more conceptual course than it was five years ago and I believe we're teaching a pretty good course. Not every teacher uses a computer algebra system, but workshops on using the computer to graph are required of all calculus students and *Maple* is available to everyone. [14] Finally, there has been and continues to be a great deal of discussion about calculus: How should we be teaching it? What should we be teaching? How do students learn mathematics? What do the students expect to learn and what do we think is important for them to learn? This ongoing discussion of difficult questions has been enlightening to each of us and continues to enrich the department.

7 Summary and Conclusions

We are still working at this, but there are, perhaps, some general conclusions that can be drawn from our project. I'll list four and comment briefly on each.

- The idea that students can learn more and better by using technology has been verified.

  Many caveats/warnings should decorate this first point (particularly the use of the word can). But all in all, I feel rather secure drawing this conclusion.

- By approaching mathematical ideas from different points of view, one exposes weaknesses in student understanding.

  These weaknesses are found with sophisticated notions such as "function" but also in more elementary notions such as "linear" or "slope" or "graph." Even our best students have deep gaps in their understanding of fundamental mathematical ideas.

- There are formidable obstacles to making any change in calculus from within the mathematics community.

  Students come to our calculus classes with expectations of us. We are practiced in meeting those expectations. The status quo is a familiar environment for both faculty and students and changing it in a significant way demands a sizable investment in time, energy, and money. Changes fostered from outside the community might meet fewer obstacles but be devastating to our educational mission [15]

- Many mathematics faculty are deeply resistant to change.

  Moreover, the reasons for this are significant and diffuse. It has been difficult to focus on what is meant by "conceptual." For some, it is synonymous with "abstract," for others it smacks of anarchy. It has been difficult, indeed at times impossible, to relegate technology to its proper subservient role. Technology does play an important role but only a supportive role. Many faculty have no knowledge of computers and have a fear of incorporating them into their teaching. Many faculty feel that the time and energy commitment necessary to change calculus are counter to their own professional interest. Indeed, often these are not the activities for which one gets promoted. Finally, changing a calculus curriculum demands faculty cooperation on a scale not many departments can accommodate.

---

[13]Of course, they also do much more than would ever be used in a calculus course.

[14]*Mathematica* is also available to all students, but on a very limited number of machines.

[15]For example, it is not inconceivable to me that schools of engineering will require only a little formal calculus training of their graduates.

Review of Calculus Lab Materials

Anita E. Solow
Grinell College

1 Philosophy of Lab Manuals

The textbook market is being flooded with lab manuals for calculus. This recent phenomena is both welcome and problematic. It is certainly evidence of the vitality of the movement to change the way calculus is taught. In particular using labs in calculus goes a long way to making our students less passive. The problem with the number of these manuals on the market is that an instructor deciding to use labs is confronted with what can seem to be an overwhelming array of choices. Hopefully this article will enable instructors to concentrate on those manuals that make the most sense for their situations. The books chosen for review in this article come from the bibliography developed by Arnold Ostebee that appeared in the August 1991 issue of *UME Trends*. Although my intention is to focus on lab manuals, some of the books referred to in that article are not lab manuals. I will offer comments on many of them anyway because the marketing of these books can be misleading. (Some books listed in that article are missing from this review because the publishers did not send the book when asked.)

Before I start reviewing the manuals that have been published, I want to set the stage by discussing what I think makes successful laboratory materials. Calculus labs can mean different things in different settings. In some cases the class goes to a special computer-equipped room and goes through the lab exercises there. In other situations, students are assigned lab projects to do outside of the classroom in an independent fashion. Sometimes these are done using computers, sometimes using graphics calculators, and sometimes working with pencil and paper. Some students write lab reports; others hand in data sheets from the lab. All of these activities can be referred to as calculus labs, and I will not differentiate among

them in this review. The question is: What makes a successful calculus lab? I believe that the answer has a great deal to do with stressing the ideas of calculus and not the wonders of technology. A good lab experience should be one where the students are able to draw conclusions based on the evidence that they have gathered. The students should be able to go through the lab questions and learn. Mindless pushing of keys should be minimized. Whether the lab is devoted to a concept of calculus or an application of calculus, the students should feel as though they accomplished something important by the time the lab is finished. There should be a sense of pride that they figured out an idea of mathematics or a problem using mathematics without being told every part of it. Students should get a feeling that they can do mathematics.

Some manuals definitely foster this feeling; others do not. In some ways it is easier to say what doesn't work than what does. The poorest manuals are the ones that basically tell the student to push this button and then that one, and never ask the student to think. Another common problem is the "Wow, Gee Whiz!," mentality that can occur in these books when the authors are so impressed with the technology and how it can be used to perform the operations of calculus, that they base their entire book on this idea. The authors know how tedious these operations would be to perform by hand, so they explain how wonderful it is now that technology is used. There are several problems with this emphasis. First, the students cannot compare the new way with the old if they have not had the experience of performing the operations by hand. They are simply not impressed. One should never forget that using a computer can be just as boring and mindless as doing pencil and paper calculations. Perhaps even more important, we are supposed to be teaching calculus, and the technology is supposed to help us do this. We should not be teaching a course in how to use technology to perform complicated, but uninteresting, calculations.

This review will be organized by software package. The main reason for this decision is that most of us are constrained by existing conditions or by financial resources in our choice of computing facilities. The hardware that we currently have or can afford will limit our choices of software. It also seems inappropriate to compare manuals that are based on different software systems if the software affects the type of questions that can be asked. The books to be reviewed will be organized into one of the following groups: *Maple, Derive, Mathematica*, Graphing Calculators, Unspecified Software, and Others (True Basic and MathCAD).

2   *Maple*

Bauldry, William C. and Joseph R. Fiedler, *Calculus Laboratories with Maple*, Brooks/Cole Publishing Company, 1991.

Geddes, Keith O., Beverly J. Marshman, Ian J. McGee, Peter J. Ponzo, Bruce W. Char, *Maple: Calculus Workbook, Problems and Solutions*, Faculty of Mathematics, University of Waterloo, 1988.

Bauldry and Fiedler's *Calculus Laboratories with Maple: A Tool, Not an Oracle*, is a fine book of labs. It contains 21 labs covering topics in Calculus I and II. Some of the more unusual topics are Euler's Method and Sky Diving, The Catenary, Summation and Inductive Verification, and Difference Equations. The problems are well presented and easy to follow, but the questions require some real thinking on the part of the students. There is a nice mix of concept-oriented labs and application labs. The *Maple* commands are given when needed, including some programs, but the emphasis is clearly on the mathematics. I recommend this book.

The book by Geddes, *et. al.*, *Maple: Calculus Workbook, Problems and Solutions*, is a collection of problems in calculus with their solutions using *Maple*. The stress is on the symbolic uses of *Maple* over the graphing. I am sure that this emphasis was partly due to the primitive state of *Maple* graphics at the time the book was written. (These graphics are quite good in *Maple V*.) The problems are varied and interesting, but are not written as labs. However, some of the problems could be used as the bases of labs written by the instructor. The book includes all of the solutions to the problems by giving actual *Maple* code that would be used to solve the problems along with the *Maple* output. This book would be a good resource for instructors looking for problems that profitably use *Maple*. In fact, some of these problems appear as labs in other collections.

3   *Derive*

Gilligan, Lawrence G. and James F. Marquardt, Sr., *Calculus and the Derive Program: Experiments with the Computer*, second edition, Gilmar Publishing, 1990.

Glynn, Jerry, *Exploring Math from Algebra to Calculus with Derive*, MathWare, 1989.

Leinbach, L. Carl, *Calculus Laboratories Using Derive*, Wadsworth Publishing Company, 1991.

Gilligan and Marquardt's *Calculus and the Derive Program: Experiments with the Computer*, is a manual of how to use *Derive* to solve calculus problems. It contains very clear explanations of the *Derive* commands to use, down to the level of which key to press when. It also tells and shows the readers (by reproducing many *Derive* screens) what they will see when they press the appropriate keys. The book shows off the capabilities of *Derive* far more than giving the students a chance to explore the ideas of calculus. For example, in the lab on the relationship between a function and its derivative the authors tell us in Objective 1 that the students will see that if a derivative of a function is positive, the function is increasing, etc. The lab itself is fine, but I question why we should tell the students the punch line. I would advocate leaving it for them to discover. I also found the following error in the book. In the algorithm for computing a lower Riemann sum, the authors state that one compares the values of the function at the two endpoints of the subinterval and chooses the smaller. Unfortunately, this gives a lower sum only if the function is monotonic on the interval. On the whole, this is not a lab manual that I would recommend.

Glynn's *Exploring Math from Algebra to Calculus with Derive*, is not a lab manual at all. It is a book that shows how *Derive* can be used to explore many areas of elementary mathematics. At times it is directed towards students, at other times towards teachers. It is written in an extremely friendly, chatty style.

The *Derive* commands are given in detail, but do not get in the way of the exposition. Although this book is definitely not a calculus lab manual, it may be useful for the novice *Derive* explorer.

Leinbach's *Calculus Laboratories Using Derive,* is a well written collection of labs for calculus that uses *Derive.* Although the *Derive* instructions are given, the book contains mostly mathematics, not computer code. The problems are interesting and varied, with several applications labs given. It is worth comparing the approach in Leinbach with Gilligan and Marquardt on the topic of the relationship between a function and its derivative. Whereas Gilligan and Marquardt give away the conclusion and leave the student to verify that, once again, the book is correct, Leinbach sets up the situation so that the students can easily draw the correct conclusions. This is what labs should be all about. The book contains twenty labs covering topics from graphing functions to functions of more than one variable.

4   *Mathematica*

Crooke, Philip and John Ratcliffe, *A Guidebook to Calculus with Mathematica,* Wadsworth Publishing Company, 1991.

Gray, Theodore W. and Jerry Glynn, *Exploring Mathematics with Mathematica,* Addison-Wesley, 1991.

Höft, Margret, *Laboratories for Calculus I using Mathematica,* Addison-Wesley, to appear early 1992. (manuscript)

Porta, Horacio and J. Jerry Uhl, *Calculus & Mathematica,* Addison-Wesley, 1991 (disks only).

Wagon, Stan, *Mathematica in Action,* W.H. Freeman and Company, 1991.

Crooke and Ratcliffe's *A Guidebook to Calculus with Mathematica,* is aptly named. It is not, however, a laboratory manual. The book goes through the basic syllabus in calculus and shows how to use *Mathematica* to solve routine calculus problems. (For some of the exercises, *Mathematica* is not needed. For others, it is useful due to the complexity of the answers.) This book is a reasonable way to learn to use *Mathematica* to solve calculus problems, if that is your goal.

Höft's *Laboratories for Calculus I using Mathematica* is one of the very few lab manuals on the market

designed for *Mathematica.* I know that many more either will be available soon or are available now on an informal basis from the developers. As the title suggests, this lab manual only covers the topics in first semester calculus. The preliminary version that I read contains ten labs, with four more planned but not yet written. Each lab consists of a *Mathematica* notebook that the student first goes through, followed by exercises that extend the ideas in the notebook. The exercises are followed by homework problems that continue the investigation of the lab and one or more challenge problems. The latter are extra-credit problems that are written in a style that would lead to more independent work than the labs themselves. In this book, the instructions given in the labs are quite specific and lead the student to fill in the blanks. Also included are Lab Report Pages that form the beginning of the lab report for each lab. These include specific instructions to the student as to exactly what should be handed in. This book gives the student more detailed information and instructions than I think it should. Because of the fixed structure of the labs, one wonders if the students would feel free to go off and explore the ideas on their own. One major improvement that is needed is a companion book for the rest of Calculus II and, possibly, Calculus III.

Calculus & Mathematica, by Porta and Uhl, is a set of electronic *Mathematica* notebooks that are a calculus course to be taught using *Mathematica.* The notebooks contain text and live calculations covering single and multivariable calculus. These are written in a problem and solution format, with the student going through the problems and solutions carefully, and then solving other problems. Details of this course can be found in *Priming the Calculus Pump* (see Section 8 for a brief discussion of this book). Although this is definitely not a lab manual, *per se,* and involves a departure from the standard classroom, it would be a worthwhile source of problems and ideas for anyone planning on using *Mathematica* in calculus, even if in a more traditional setting.

The books by Gray and Glynn and by Wagon are about exploring mathematics with *Mathematica.* The difference between the books is in the style of the presentation and the sophistication of the mathematics. *Exploring Mathematics with Mathematica,* by Gray and Glynn, is written as a series of dialogues between the authors that the reader is invited to join. This book is extremely friendly and would be fun to go through, with the reader becoming a participant along with the authors. A great deal of good mathematics is visited along the way. One unusual feature of this book is that it comes with a disk for use on a CD-ROM player. The disk contains the entire book and *Mathematica* code to allow the user/reader to continue to play with and explore the ideas. *Mathematica in Action*, by Stan Wagon, covers substantially higher-level mathematics and is written in a more traditional manner. I recommend both of these books, but not as lab manuals.

5   Graphics Calculators

Beckmann, Charlene E. and Ted Sundstrom, *Graphing Calculator Laboratory Manual for Calculus,* Addison-Wesley, 1991. (Second edition available in 1992.)

Minton, Robert B. and Robert T. Smith, *Discovering Calculus with the HP-28 and the HP-48,* McGraw-Hill, 1992.

Others not reviewed:

1. J. Eidswick and J. C. Matthews, *Calculus Companion for the HP-28S and HP-48SX,* Harper Collins, 1991.

2. G. Foley, *Calculus with TI-81 Graphing Calculators,* Wadsworth Publishing, 1991.

3. L. E. Garner, *Calculus with the Hewlett Packard Symbolic Manipulating Calculators,* Dellen Publishing Co., 1991.

4. J. Kennelly and J. H. Nicholson, *Calculator Enhancement for a Course in Single-Variable Calculus,* Harcourt Brace Jovanovich, 1991.

5. D. Pence, *Calculator Activities for Graphic Calculators,* PWS-Kent, 1990.

6. J. A. Renecke, *Calculator Enhancement for a Course in Multivariable Calculus,* Harcourt Brace Jovanovich, 1991.

The book by Beckmann and Sundstrom is an excellent lab manual for calculus using graphics calculators. I found this collection of labs to be both realistic in its expectations of what students can accomplish and challenging enough to ask the students to think. This book is clearly written. The instructions for the calculators are carefully given in the back section of the book. This way, the labs are about mathematics, not about which button to push. The authors

have organized the book very cleverly so that the labs are first, followed by a section for the Casio Graphics Calculator and then one for the TI-81. These include short introductions to the calculators, a list of all of the labs and which program is needed for each, followed by the code (clearly annotated) for the programs. I admit that I prefer working with computers than calculators, but I understand the economic advantage and the convenience associated with the use of calculators. I used the TI-81 when I went through this collection. The labs were interesting and well written and not too difficult to get the calculator to supply the necessary information.

Minton and Smith have written a book about using the HP supercalculators for calculus. The book is a collection of chapters, each organized about a central idea of calculus. Too much of the book concerns how to use the HP calculators. The first fifty-four pages of the book are devoted to the workings of the HP. Subsequent chapters contain many programs. I appreciated the TI-81 a great deal more after dealing with programming the HP. Because HP programming is much like assembly language programming, it is complex enough to require a great deal of discussion. Each section of the book begins with a quick summary of the calculus topic, with numerous examples worked on the HP. This is followed by a set of standard exercises. Finally, each section ends with an Exploratory Exercise. These latter problems often concern an application or extension of the mathematics. Some are interesting, but this is not a lab manual. With some work, several of the Exploratory Exercises and regular exercises could make the basis for labs.

6   Unspecified Software

Evans, Benny and Jerry Johnson, *Uses of Technology in the Mathematics Curriculum,* Mathematics Department, Oklahoma State University, 1990.

Moody, Michael E. and Kevin Shannon, *Microcomputer Exercises for Calculus: A Laboratory Manual,Microcomputer Exercises for Calculus: A Laboratory Manual,* West Publishing Company, 1988.

Small, Donald B. and John M. Hosack, *Explorations in Calculus with Computer Algebra Systems,* McGraw Hill, 1991.

Solow, Anita E., ed., *Learning by Discovery,* available from Professor Wayne Roberts, Department of Mathematics, Macalester College, St. Paul, MN 55105.

Evans and Johnson's book starts with an introduction that I wish I had written. They emphasize that "this is not a computer manual but a mathematics manual." Their two guiding principles are "Our business is Mathematics; we are not interested in the computer, *per se.*" and "Passivity is to be avoided." This book does not contain particular software commands in the labs, nor does it provide details about adapting the labs to the most commonly used software packages. The labs are varied and plentiful and cover precalculus, calculus, and linear algebra. Most of these labs are actually outlines and will require some local adaptation before being handed out to students. The labs vary considerably in difficulty. Some in the beginning are quite short and not very ambitious. This makes sense while the students are getting used to using the computer. Some of the later labs have more substance. For example, the brachistochrone lab is quite delightful and challenging. I found some of the labs on multivariable calculus disappointing. One showed how to use the computer to do the algebra to find critical points of a function of two variables. There was, unfortunately, no mention of the geometry of the situation. On the whole, though, there are many good labs in this book. Moreover, each lab also includes an instructor sheet that contains complete answers to the problems in the lab. One very helpful section is a review of the existing software packages and their cost.

Moody and Shannon have written an interesting lab manual. Some unusual titles include "A Journey through the Center of the Earth," "Alternating Currents," "The Distribution of IQ Scores," "Halley's Comet," and "Computer Spirograph." The book also includes some labs on Calculus III topics (e.g., partial derivatives and 3-dimensional surfaces). I am disappointed that Moody and Shannon have taken an innovative and creative collection of labs and have placed it in a highly restrictive format. Each lab is composed of three pieces. First is a Checklist that spells out what you need for the lab in terms of software and mathematical background. It then lists the steps that the student needs to take to complete the lab and spells out exactly what is to be handed in. This is followed by the Exercise itself. This gives the context for the lab and the physical situation that is being analyzed. The labs are written in a very open-ended style, which, without the supporting materials, would make the basis for advanced independent student work. The last part of the lab is the Worksheet, which is a fill-in-the-blank formatted page or two that

the student is asked to fill, in order to complete the lab. It tells the students, step by step, exactly what to do in order to solve the problem that was posed in the Exercise part. The Worksheet provides the students with the answers to how to attack the problem, leaving little for the students to do other than compute (or use the computer to compute) the specific answers to the small questions. The authors also provide a symbol next to each question that tells the students whether to use pencil, calculator, computer, or an idea to solve that question. I do not care for giving students this much instruction on how to solve a problem, although the authors use these labs in a situation where the students do all their labs independently.

Small and Hosack's book, *Explorations in Calculus with Computer Algebra Systems,* differs from the other books reviewed in this section in that the book uses *Maple* syntax throughout, but gives an Appendix with information about how to use several other software packages, including lists of the basic commands in each package. Each section starts with some mathematical discussion, followed by several worked examples, ending with a set of exercises. Most of these exercises are different from those found in typical textbooks and contain phrases like "What do you think ...?" There is a great deal of material in this book, starting with graphing and ending with functions of several variables. This book is more of a supplement to calculus than a lab manual, although much of it could be adapted to the lab format if desired. In fact, several problems from this book have appeared as labs in other collections.

Learning by Discovery, a collection of labs edited by this author, is written without any computer or calculator instructions in the body of the lab. However, it contains instructions and programs for using the labs with *Maple, Derive,* and *Mathematica.* Many of the labs can also be done successfully with a graphics calculator. These labs stress the concepts of calculus and are based on the assumption that the students can draw valid conclusions about the way calculus works from well developed examples. Like Evan and Johnson's book, the labs themselves do not contain computer commands. However, unlike that book, most of the information for adapting the labs is found in the Notes to the Instructor that follows each lab. There are twenty-six labs covering topics from all three semesters of calculus. This, of course, is my favorite collection of lab materials.

7 Other Packages

Rowell, James W., *Mathematical Modeling with Math-CAD: Explorations in the Calculus and Points Beyond,* Addison-Wesley, 1989.

Miech, R. J., *Calculus with MathCAD,* Wadsworth Publishing, 1991.

Hurley, James F. and Charles W. Paskewitz, *Computer Laboratory Manual for Calculus,* Wadsworth, 1988. This is designed to be used with True BASIC.

I am less familiar with MathCad and True Basic than I am with the other packages discussed earlier. Therefore, my comments about these books are based on my reading of the books, but not on my actually using the books in conjunction with the required software package.

Rowell's book, *Mathematical Modeling with Math-CAD: Explorations in the Calculus and Points Beyond,* consists of a series of Worksheets and Assignments. The topics include differential and integral calculus, linear algebra, topics in calculus, and special topics. The worksheets are actual MathCAD output and are also supplied on a disk with the book. They consist of background information and worked examples. The user is supposed to read this and then use the disk to work through the examples and to experiment. The assignments mainly ask the students to modify the worksheets to perform similar calculations. At the end of the assignment sheets there are Challenge Problems and New Directions. These are more open-ended problems for the students to investigate. Some of these pose genuinely difficult and interesting problems.

Miech's book is a tutorial on MathCAD and its uses in calculus. The purpose is to demonstrate how to use MathCAD to perform relatively routine calculus problems. As such, it may be useful for some students who wish to learn to use MathCAD on their own, but I do not see the appeal of this book as part of a calculus course.

Hurley and Paskewitz's book is devoted to True BASIC. The thrust of the book is to have students type in computer programs that are developed and printed in the book. These programs perform standard calculus procedures such as finding extreme values, curve sketching, etc. Occasionally students are asked to modify a given program in an exercise at the end of the chapter. Usually these exercises are designed to have the student apply the program given in the chapter to specific functions. This is the only computer oriented book to be reviewed in this article that takes a programming approach to calculus labs, but the approach it takes does not require a great deal of programming.

Although I find this particular book to be unattractive, there is a legitimate philosophical question here: Should students use software packages or should they write programs in a computer language as a means of learning calculus? I would argue that, ideally, the software packages are a more appropriate vehicle. I would like my students to focus on the mathematics and not on the programming language. Of course, I would prefer software packages that are easy to use and do not require learning a foreign syntax. Proponents of programming claim that students get a thorough understanding of the subject if they have to program the instructions into the computer. Actually, given the current state of computer algebra systems, these two positions are not as far apart as they seem. Often the syntax required to use a software package can be a formidable barrier to use, and most computer algebra systems allow the user to write programs within the language.

8 Other Resources

Ellis, Wade, Jr. and Ed Lodi,

> *Maple for the Calculus Student,* Brooks/Cole, 1989.
>
> *A Tutorial Introduction to Mathematica,* Brooks/Cole, 1991.
>
> *A Tutorial Introduction to Derive,* Brooks/Cole, 1991.

Leinbach, L. Carl, ed., *The Laboratory Approach to Teaching Calculus,* Mathematical Association of America, 1991.

Tucker, Thomas, W., ed., *Priming the Calculus Pump: Innovation and Resources,* Mathematical Association of America, 1990.

The three books by Ellis and Lodi are basically the same. They carefully and gently introduce the student to the features of the software that the students would need for calculus (and linear algebra). These books are good references for the students since they are so easy to read and follow; they do not go into any of the fancier features of the particular packages. Instructors will quickly find topics that the books do not cover, but it should take most students much longer to discover that limitation. The best way to think about these books is that they are like security blankets for the students. The books should help ease the students' fears about the software package being used, but they probably are not needed. At least they do no harm.

The MAA Notes edited by Leinbach contains examples from twenty-seven institutions that use the lab approach in calculus. It is a gold mine of ideas for labs, both what works and what doesn't. It does not contain too many fully developed labs, however. Most of the articles give examples from the labs that the authors use. Therefore, the material in this book will help the instructor to develop labs, but little could be handed directly to the student. This book is definitely worth owning and reading if you are interested in the lab approach to calculus.

The MAA Notes edited by Tucker is a handy guide to what is going on across the country in calculus reform. Several projects are examined in depth; others are briefly described. Some of the projects use calculus labs; others do not. Several of the articles are sufficiently detailed to contain large portions of labs, exercises, and exams. The book also contains information about whom to contact in the different programs, what software is available to aid in mathematics teaching, and a description of the features and price of various graphics calculators. This book is an extremely valuable resource for finding out the current (as of publication) state of affairs.

9 New Lab Manuals

There are several laboratory manuals that are scheduled to be published soon. Although I have yet to read these books, you may find it useful to know what will be available in the near future.

D. C. Arney, *Exploring Calculus with Derive,* Addison-Wesley, 1992.

B. Braden, D. Krug, S. Wilkinson, and P. McCartney, *Problems Manual for Calculus,* John Wiley & Sons, 1992.

R. Decker and J. Williams, *Calculus Laboratory Manual,* Prentice-Hall, 1992.

B. Evans and J. Johnson, *Derive Problems Manual of*

Calculus, John Wiley & Sons, 1992.

K. Harris, *Maple Problems Manual for Calculus,* John Wiley & Sons, 1992.

M. Lehmann and J. Finch, *Exploring Calculus with Mathematica,* Addison-Wesley, 1992.

J. McCarter, *Graphing Calculator Manual for Calculus,* John Wiley & Sons, 1992.

Labs are being done in calculus all over the country, from large universities to small colleges. There is far more material than has been published and/or reported on in this article. Now that a great deal of material exists, it is time to use it; we need not all create our own labs from scratch.

Good luck in your labs; I wish you and your students happy exploring.

Computer Algebra in Undergraduate Mathematics: An Annotated Bibliography

Stanley E. Seltzer
Ithaca College

1   Introduction

As mathematics instructors have incorporated computer algebra systems (CAS) in their courses, they have begun to present conference papers, publish journal articles, and write books related to this subject. These works may describe a sample class or laboratory experience, comment on the impact on a course, report on a formal experiment, or discuss general issues. This bibliography provides a sampling of the literature that has appeared since 1986; it consists of short entries for a number of papers, articles, proceedings volumes, and books. In most cases I have given a brief description of the article (book, volume); it is not my intention to provide a "review" of every item. This is by no means a complete survey of the literature. Among the things that are not included are articles that describe pre-college use of computers, articles involving non-CAS use of computers in undergraduate mathematics, and items that I was unable to locate. The bibliography begins with several conference proceedings and collections, followed by a list of books and a list of articles.

2   Proceedings and Collections

Three of these (two MAA Notes volumes and the proceedings of the 1985 ICMI Symposium) are so rich in useful articles that I did not list any of the individual papers below; you will find valuable reading in each of these volumes.

• Banchoff, T. F. *et al.*, eds. 1988. *Educational Computing in Mathematics*. North-Holland.

This includes all invited talks and selected contributed papers and software demos for the International Conference ECM/87 titled "Educational Computing in Mathematics" held in Rome in 1987. It includes the papers by Flanders and Koçak that are listed below.

• Cooney, T. J. and Hirsch, C. R., eds. 1990. *Teaching and Learning Mathematics in the 1990s*. National Council of Teachers of Mathematics.

This 1990 yearbook addressing changing roles to teachers and students in the face of calls for reform includes the papers by Demana and Waits and Heid, Sheets, and Matras that are listed below.

• Demana, F., Waits, B. K., and Harvey, J., eds. 1990. *Proceedings of the Second Annual Conference on Technology in Collegiate Mathematics*. Addison-Wesley.

This volume is from the Annual Conference on Technology in Collegiate Mathematics, held in Columbus, Ohio, in November 1989. It includes papers by Beckmann, Dick, Goldenberg, Harvey and Osborne, Johnson and Lamagna, and Steen, listed below. The conference organizers have indicated that the proceedings of the 1991 conference will be published by Addison-Wesley.

• Howson, A. G. and Kahane, J. P., eds. 1986. *The Influence of Computers and Informatics on Mathematics and its Teaching*. Cambridge University Press.

This volume contains a report of the 1985 ICMI Symposium held in Strasbourg and eleven selected papers. (Supporting papers are published in a separate volume *Supporting papers of the ICMI Symposium*, IREM, Université Louis-Pasteur, Strasbourg, 1985, which I couldn't locate.) The three parts of the 35-

page report are titled "The effect on mathematics," "The impact of computers and computer science on the mathematics curriculum," and "Computers as an aid for teaching and learning mathematics." Rather than reviewing some or all of the eleven papers, I encourage you to read the report and the papers that interest you.

● Leinbach, L. C. *et al.*, eds. 1991. *The Laboratory Approach to Teaching Calculus*, MAA Notes Number 20, Mathematical Association of America.

As the title suggests, this volume is full of material related to calculus. Some of the papers in Part I, "General issues related to the laboratory approach," apply more generally, while Parts II and III, "Examples of established calculus courses with laboratories" and "Projects for use in calculus laboratories," are calculus specific. The collection is quite broad: the courses described use a variety of hardware and software platforms (and some non-computer labs), and use of the laboratory includes closed labs, extended laboratory projects, and classroom/labs.

● Smith, D. A. *et al.*, eds. 1988. *Computers and Mathematics: The Use of Computers in Undergraduate Instruction.* MAA Notes Number 9, Mathematical Association of America.

This is an excellent source for anyone who is interested in what can be done with CAS in the mathematics curriculum. A project of the MAA Committee on Computers in Mathematics Education (CCIME), this volume consists of nineteen papers including a short historical preface and an interesting general survey (Alan Schoenfeld's "Use of Computers in Mathematics Instruction"); essays on computer algebra systems, evaluating mathematical software, and differential equations software; two approaches to calculus; and papers discussing use of the computer in discrete mathematics, linear algebra, differential equations, statistics, probability, "liberal arts mathematics," geometry, abstract algebra and number theory, remedial mathematics courses, and "transition" mathematics courses.

3 Books

There seem to be relatively few books so far that do not fall into the category of calculus lab manual. Many of these are reviewed by Anita Solow in her "Review of Calculus Lab Materials," so this is a short section.

● Akritas, A. G. 1989. *Elements of Computer Algebra.* John Wiley.

This is a text, complete with exercises and notes following each chapter, that grew out of the author's senior/graduate Computer Algebra course at University of Kansas and National Technical University of Athens, Greece. Part I is an introduction; Part II addresses mathematical foundations and basic algorithms; and Part III deals with applications and advanced topics including error-correcting codes and cryptography, computation of polynomial gcd's and polynomial remainder sequences; factorization of polynomials with integer coefficients; isolation and approximation of the real roots of polynomial equations.

● Davenport, J. H., Siret, Y., and Tournier, E. 1988. *Computer Algebra: Systems and Algorithms for Algebraic Computation.* Academic Press.

This is a book for the reader who wants to know "how does computer algebra work?" The authors believe that "it is not sufficiently well-known that [symbolic computation] is as much within the range of computers as is numerical calculation," and have written this book "to demonstrate the existing possibilities, to show how to use them, and to indicate the principles on which systems of non numerical calculation are based, to show the difficulties which the designers of these systems had to solve—difficulties with which the user is soon faced."

The first chapter introduces computer algebra through a series of examples, mostly in MACSYMA. Subsequent chapters deal with the problem of data representation; polynomial simplification; advanced algorithms; and formal integration and differential equations. This book is heavily mathematical, and there is a fifteen-page bibliography for the reader who wants to learn more.

● Hubbard, J. H. and West, B. H. 1991. *Differential Equations: A Dynamical Systems Approach. Part I: Ordinary Differential Equations.* Springer-Verlag.

This is a significant text in that it really acknowledges that the computer changes the way you teach differential equations. As the authors say in the preface: "The origin of this book, and of the programs (which preceded it) was the comment made by a student in 1980: 'This equation has no solutions.' The equation in question did indeed have solutions as an immediate consequence of the existence and uniqueness theorem, which the class had been studying the previous month. What the student meant was that there were no solutions that could be written in terms of elementary functions, which were the only ones he believed in. We decided at that time that it should be possible to use computers to **show** students the solutions to a differential equation and how they behave, by using computer graphics and numerical methods to produce pictures for qualitative study. This book and the accompanying programs are the result."

The accompanying programs, MacMath, include Analyzer, Diff Eq, Num Meths, Cascade, and 1D Periodic Equations. Part I of the text covers ODE's with chapters titled "Qualitative methods," "Analytic methods," "Numerical methods," "Fundamental inequality, existence, and uniqueness," "Iteration," and an appendix "Asymptotic development." Part II covers systems of ODE's, Part III PDE's.

4   Papers

• Aspetsberger, K. and Kutzler, B. 1988. Symbolic computation—a new chance for education. In F. Lovis and E.D. Tagg, eds., *Computers in Education*, North-Holland, 331–336.

This paper gives a brief discussion of four applications of symbolic computation giving an annotated example of each: computer algebra (using muMATH to solve an optimization problem and to simplify expressions), computational geometry (specifying geometric objects), automatic reasoning (using PROOF-PAD to prove that the sum of bounded functions is bounded), and automatic programming (writing a sort routine in PROLOG). The authors believe that in all four cases, the computer can automate tedious tasks, improve individual training, motivate algorithmic problem solving, and improve the capability of abstraction.

• Ayers, T., Davis, G., Dubinsky, E., and Lewin, P. 1988. Computer experiences in learning composition of functions. *J. for Res. in Math. Ed.* 19:246–259.

This paper reports on an experiment using Unix pipes to teach composition of functions. It includes a discussion of Piaget's notion of reflective abstraction, the conceptual basis for the authors' constructive approach to learning mathematics. The data suggests that computer experiences were more effective, although the sample sizes are quite small.

• Beckmann, C. E. 1991. Appropriate exam questions for a technology-enhanced Calculus I course. In [Demana, Waits, and Harvey, 1990], 118–121.

Use of technology has raised thorny questions re-

garding testing: Are there questions that are no longer valid? Are there questions that are now appropriate? What kinds of questions might be asked that do not disadvantage those students who do not have the technology readily available on exams? In this short paper, the author gives five examples of exam questions that assess students' "mathematical power" (as outlined by the NCTM) in a technology-enhanced course.

• Bloom, L. M., Comber, G. A., and Cross, J. M. 1986. Use of the microcomputer to teach the transformational approach to graphing functions. *Int. J. Math. Educ. Sci. Technol.* 17:115–123.

This paper reports "considerable success with senior secondary students and trainee teachers" using computer programs as aids in teaching the transformational approach to graphing functions. They describe three programs: STDFNC (which graphs the standard functions: $mx + b$, x^2, x^3, trigonometric functions, e^x, $\ln(x)$, $|x|$, $\sqrt{x}$, $[x]$, and $1/x$); TFGPH (which applies translations, reflections, and dilations to a graph and its algebraic expression); and GUESS (which presents a graph for the student to identify by using TFGPH).

• Buchberger, B. 1990. Should students learn integration rules?. *SIGSAM Bull.* 24(1):10–17.

The author begins by explaining that his title really means: should math students learn area X of mathematics when this area has been trivialized? (By "trivialized" is meant that there is a feasible or efficient or tractable algorithm that can solve any instance of a problem from this area.) He gives some examples: arithmetic on the natural numbers and integration, among others; provides extreme answers and counter-arguments on both sides; and articulates the "Black-Box/White-Box Principle," concluding that tools used to **introduce** a subject should not be used as a black box.

• Cromer, T. 1988. Linear algebra using muMATH. *Collegiate Microcomputer* 6:261–268.

The author shows how muMATH can perform many linear algebra calculations required of undergraduates; he also mentions how easily the instructor may use muMATH to construct matrices with "nice" properties (determinant 1, with specified eigenvalues, or symmetric with specified eigenvalues) for paper-and-pencil student exercises.

• Demana, F. and Waits, B. K. 1990.   Enhancing

mathematics teaching and learning through technology. In [Cooney and Hirsch, 1990], 212–222.

Based on the Calculus and Computer Precalculus Project, for which "technology" means access to graphing calculators, this paper discusses things to realize: what you can do, what you should do (that you didn't need to without technology).

• Devitt, J. S. (1989). Unleashing computer algebra on the mathematics curriculum. In *Proceedings of the ACM-SIGSAM 1989 International Symposium on Symbolic and Algebraic Computation*, ACM Press, 218–227.

Devitt describes use of symbolic algebra during lectures in a first-year calculus course, remarking that "we have been pleasantly surprised by the effectiveness of this approach." The nine examples (these range from showing that the shortest distance from a point P to the graph of a function f is along a line that is normal to the graph to deriving Simpson's rule), include some discussion of pedagogy along with the *Maple* dialogs.

• Dick, T. 1991. The epsilon-deltas of scaling—using graphing technology in calculus. In [Demana, Waits, and Harvey, 1990], 144–147.

The author shows how one can use a graphing calculator to address the rigorous definition of limit, continuity, and derivative. For example, the "scaling definition" of continuity of f at a is that given any vertical scaling for a screen centered at the pixel $(a, f(a))$, there exists a horizontal scaling such that the graph of $y = f(x)$ is horizontal.

• Fey, J. T. 1989. Technology and mathematics education: a survey of recent developments and important problems. *Educ. Studies in Math.* 20:237–272.

In this broad overview, prepared as a survey lecture at the sixth ICME in Budapest, Fey discusses six aspects of technology applied to mathematics education: numerical computation; graphic computation (which he describes as the most appealing development); symbolic computation; multiple representations of information; programming and connections of computer science and mathematics curricula; and artificial intelligence and machine tutors. The description of symbolic computation is quite brief: he feels that the "promise and potential problems that may result from [use of CAS] are largely unknown at this time." There is an extensive bibliography (five pages long).

• Flanders, H. 1988. Teaching calculus as a laboratory course. In [Banchoff *et al.*, 1988], 43–48.

This paper describes the author's use of his software package MicroCalc 3.0 in one pilot section (22 students) of Calculus 1, concluding that one should separate lecture/demonstrations from labs. He also mentions that his three objectives in designing Micro-Calc were to (1) provide a tool for experiments, (2) remove the drudgery of calculation, and (3) provide a tool for checking results. He proposes "standards" for educational software including friendliness and syntax close to mathematical.

• Foster, K. R. and Bau, H. H. 1989. Symbolic manipulation programs for the personal computer. *Science* 243:679–684.

Although dated, this may be interesting for historical purposes. The authors review *Derive* (version 1.1), *Maple* (version 4.2), *Mathematica* (version 1.1), *Reduce* (version 3.3), and MACSYMA (version 309.6). All but the latter were available on personal computers; MACSYMA was anticipated. Among their observations: the *Mathematica* interface was described as superb; MACSYMA is "somewhat difficult to use"; *Derive* has "comparatively limited capabilities" but "performs well for the problems for which it was intended." The overall conclusion is that "these programs can be productively used as sophisticated calculators, and learned in a few hours."

• Goldenberg, E. P. 1991. The difference between graphing software and educational graphing software. In [Demana, Waits, and Harvey, 1990], 34–42.

This is a very interesting paper that identifies "lessons" learned from using standard graphing software for educational purposes. A few of the lessons—some quite radical—are really suggestions for how one can use graphical software. They range from "Lesson 1: It must be easy to modify functions" to "Lesson 13: Curriculum must accommodate visual thinking."

• Harvey, J. G. and Osborne, A. 1991. Calculator and computer technologies in mathematics classrooms: a user's guide. In [Demana, Waits, and Harvey, 1990], 74–86.

This paper is "an accounting of the problems and solutions to those problems that we have encountered as technologies have become a part of instruction in colleges, universities, and high schools." Issues range from "Technologies: What You Have and What You Can Get" and "Teaching Faculty, Staff, and Students" to "Obtaining Needed Approvals," "Classroom Facilities," "Class Dynamics and Pedagogy," "Testing," and "Spreading the Word."

• Heid, M. K. 1988. Resequencing skills and concepts in applied calculus using the computer as a tool. *J. for Res. in Math. Ed.* 19:3–25.

The study described in this paper addresses two questions: (1) Can the concepts of calculus be learned without concurrent or previous mastery of the usual algorithmic skills of computing derivatives, computing integrals, or sketching curves? and (2) How does student understanding of course concepts and skills attained in a concepts-first course differ from that attained by students in a traditional version of the course? The paper includes excerpts of interviews with students from experimental and traditional sections as well as sample questions. Interestingly, the "concepts-first" course concluded with three weeks of traditional (skills-oriented) preparation for a traditional final; students from the experimental sections did almost as well as those in the traditional sections.

• Heid, M. K., Sheets, C., and Matras, M. A. 1990. Computer-enhanced algebra: new roles and challenges for teachers and students. In [Cooney and Hirsch, 1990], 194–204.

What is reported here about the authors' experiences with their ninth-grade Algebra with Computers course applies equally to all instructors who use computers in their teaching. (And much applies to those who don't.) The paper discusses roles for teachers as technical assistant, collaborator, facilitator and catalyst; responsibilities and challenges in evaluating student learning, in time allocation, in conducting and planning classroom activities; and student challenges of self-directed learning.

• Johnson, D. L. and Lamagna, E. A. 1991. The Calculus Companion: not just another pretty interface. In [Demana, Waits, and Harvey, 1990], 231–236.

The authors illustrate their Calculus Companion, a kinder and gentler CAS designed for use in calculus. (The present version is now called Newton.)

• Judson, P. T. 1990. Calculus I with computer algebra. *Journal of Computers in Mathematics and Science Teaching* 9(3):87–93.

The author describes an experimental study in Calculus I where the instructor used computer algebra in class and students were required to complete computer algebra homework assignments. These students did about as well as students in a control section (no

computer algebra); student reaction was mixed, although most of the negative comments were that the computer required too much time. The paper includes excerpts from a handout and an assignment as well as advice for the instructor who is interested in using computer algebra.

• Judson, P. T. 1990. A computer algebra laboratory for Calculus I. *Journal of Computers in Mathematics and Science Teaching* 10(4):35–40.

This paper describes a one-credit laboratory for Calculus I, including the physical setting, a brief description of the ten labs, student reaction, and advice. Judson mentions two things that I found interesting: (1) Because of scheduling difficulties, some of the students took the lab on an independent-study basis, and they did almost as well as the supervised students. (2) Students "particularly liked working as partners if they each had their own machine. In this way they could both experiment and compare their results." The author reports unqualified success with the experience, commenting along the way that "it was gratifying to see that questions changed almost immediately from how to use *Maple* to questions concerning how to do the mathematics."

• Koçak, H. 1988. Computer experiments in differential equations. In [Banchoff *et al.*, 1988], 75–91.

The author describes PHASER (an animator/simulator for difference and differential equations), which he uses as a tool for geometric study and numerical simulation in a new course in ordinary differential equations. He also proposes an outline for a new second course.

• Kwok, Y. K. 1991. Application of MACSYMA to solutions of ordinary differential equations. *Int. J. Math. Educ. Sci. Technol.* 22:877–888.

The author shows how MACSYMA can be used to solve ordinary differential equations encountered in the sophomore- or junior-level course.

• Lance, R. H., Rand, R. H., and Moon, F. C. 1986. Teaching engineering analysis using symbolic algebra and calculus, *Eng. Educ.* 76:97–101.

This paper describes a very early use of CAS by students in a 1985 sophomore engineering class at Cornell University. (This use of MACSYMA extended more limited use dating back to 1983.) A few examples are cited. The authors conclude that "we are convinced that engineers in the future will use both numerical and symbolic computation and now is the time to start training them accordingly."

• Mathews, J. H. 1989. Computer symbolic algebra applied to the convergence testing of infinite series. *Collegiate Microcomputer* 7:171–176.

The author shows how muMATH can perform the calculations needed to perform the nth term test, ratio test, comparison test, integral test, and Raabe's test.

• Mathews, J. H. 1990. Teaching Riemann sums using computer symbolic algebra systems. *College Math. J.* 21:51–55.

The author shows how *Maple* can be used to compute Riemann sums.

• Mathews, J. H. 1991. *Mathematica*, a new tool for studying optimization. *Int. J. Math. Educ. Sci. Technol.* 22:569–576.

The author shows how *Mathematica* can plot functions, compute and find zeros of (first and second) derivatives, etc. He concludes that "we now have the opportunity to start developing new course materials and to incorporate this emerging technology into the mathematics curriculum."

• Nievergelt, Y. 1987. The chip with the college education: the HP-28C. *Amer. Math. Monthly* 94:895–902.

The author demonstrates the HP-28C (then quite new), illustrating its utility in solving standard calculus, statistics, numerical analysis, and linear algebra problems and introducing some of its programming capabilities. He also reports reactions from several mathematics faculty, employers of mathematicians, and executives. Among the conclusions in this thoughtful review: "It appears that a new trend toward the use of the HP-28C and its successors would require that students understand the underlying concepts even better than before in order to decide what computations to perform, to interpret the results with lucidity, or even first to recognize that no calculator can address the issue at hand."

• Page, W. 1990. Computers algebra systems: issues and inquiries. *Computers Math. Applic.* 19:51–69.

This is a wide-ranging article that raises a number of issues well worth considering, among them (1) How can CASs be used to provide students with greater understanding and deeper insights than heretofore possible? and (2) What new awareness and knowledge do **we** need in order to effectively harness and realize CASs' great potential for educational gain? The author is clearly bullish on CASs, but warns us to be aware of the potential for their misuse. Page observes that if we wish to stress concepts, our tests ought to be " 'define and illustrate', 'explain', 'express', 'describe', 'compare', ... not 'solve', 'sketch', 'evaluate', ..."

• Shumway, R. 1990. Supercalculators and the curriculum. *For the learning of Math.*, 10:2–9.

In this paper the author shows how to use a "personalized" HP calculator to investigate several problems. (By "personalized," he means that the calculator contains programs that encapsulate useful operations, like zooming in on a graph.) For example, one problem is to analyze the function $\cos(x) + \sin(100x)/100$; another is to solve $\ln(x+1) + \ln(x-1) = 3$. He mentions some of the mathematical issues that these problems raise, among them the relationship between a function and its derivative (in the first problem) and symbolic, numerical, and graphical approaches to the second. The paper concludes with some discussion of relevant research.

• Small, D., Hosack, J., and Lane, K. 1986. Computer algebra systems in undergraduate instruction. *Coll. Math. J.* 17:423–433.

This is a very early paper that introduces computer algebra systems and describes their use in calculus. There is an extended discussion of a classroom approach to graphing rational functions and shorter treatment of critical points and extrema, numerical integration, power series, and Lagrange multipliers. I consider this a classic in some respects—it was the first paper about CAS in the classroom that I ever saw.

• Steen, L. A. 1991. Twenty questions for computer reformers. keynote address at the 1989 Conference on Technology in Collegiate Mathematics. In [Demana, Waits, and Harvey, 1990], 16–19.

This is a summary of Steen's keynote address. The paper includes a paragraph on each of the twenty questions, which range from "Can computers help students understand mathematics?", "Can students develop mathematical intuition without performing extensive mathematical manipulations?", and "Will using computers reduce students' facility to compute by hand?" to "How does computing change what students should know about mathematics?", "How much time and distraction is computing worth?", "How much programming should be taught in mathematics courses?", and "If computers handle routine calculations, what will students do instead?'"

● Thomas, D. A. 1990. Giving meaning to matrices with MATLAB. *Journal of Computers in Mathematics and Science Teaching* 9(3):73–85.

This paper gives a short description of MATLAB and gives two applications of matrices: reflecting a point about a line (in the plane) and roll, pitch, and yaw. The author argues that the teaching of matrix algebra can be enriched by a computer environment.

● White, J. E. 1988. Teaching with CAL: a mathematics teaching and learning environment. *College Math. J.* 19:424–443.

In the context of his CAL system, the author examines the advantages and applications of symbolic (in contrast to numeric) computing in teaching mathematics. The paper is rich in examples (among them Taylor polynomials, sums of powers, structural stability of plane maps, iterated maps of the interval, parallel translation in spherical geometry) illustrating use of the computer as a "smart blackboard" and as a laboratory tool.

● White, J. E. 1989. Mathematics teaching and learning environments come of age: some new solutions to some old problems. *Collegiate Microcomputer* 7:203–224.

White sees that the greatest promise of the computer is that it can provide a teaching/learning tool that allows students to explore mathematics in groups or on their own. He writes thoughtfully on problems associated with using teaching/learning environments, among them the fact that most CASs are not designed specifically for the classroom. Much of this article is devoted to the author's CAL system.

● Wieder, S. 1991. Using MathCAD in general calculus. *Journal of Computers in Mathematics and Science Teaching* 10(4):57–63.

Wieder describes how MathCAD can be used for demonstrations in a calculus course, giving three examples: minimizing the cost of fence bounding a rectangle of fixed area, investigation of convergence of p-series and geometric series, and computing the area between two polar curves.

● Zorn, P. 1987. Computing in undergraduate mathematics. *Notices Amer. Math. Soc.* 34:917–923.

This "issues paper" was written in conjunction with a conference that examined the role of computing in undergraduate mathematics. It is interesting to keep in mind the situation at that time: " 'computer al-gebra systems' (*Macsyma, Maple, Reduce*, SMP, and others) are starting to appear on student-type machines." The author writes thoughtfully about mathematics and computing, the past and future, and benefits and opportunities. The paper concludes with open questions, among them "How should analytic and numerical viewpoints be balanced?", "How does computing change what students should know?", "How will computing affect advanced courses?"

5   Other sources

There are several other sources of material on uses of computer algebra systems in undergraduate mathematics that are not represented here. The *Notices of the American Mathematical Society* has a regular column titled "Computers and Mathematics." The second half of 1992 will see the first issue of a new journal, *Journal of Technology in Mathematics*; *Mathematica Journal* began publishing in 1991.